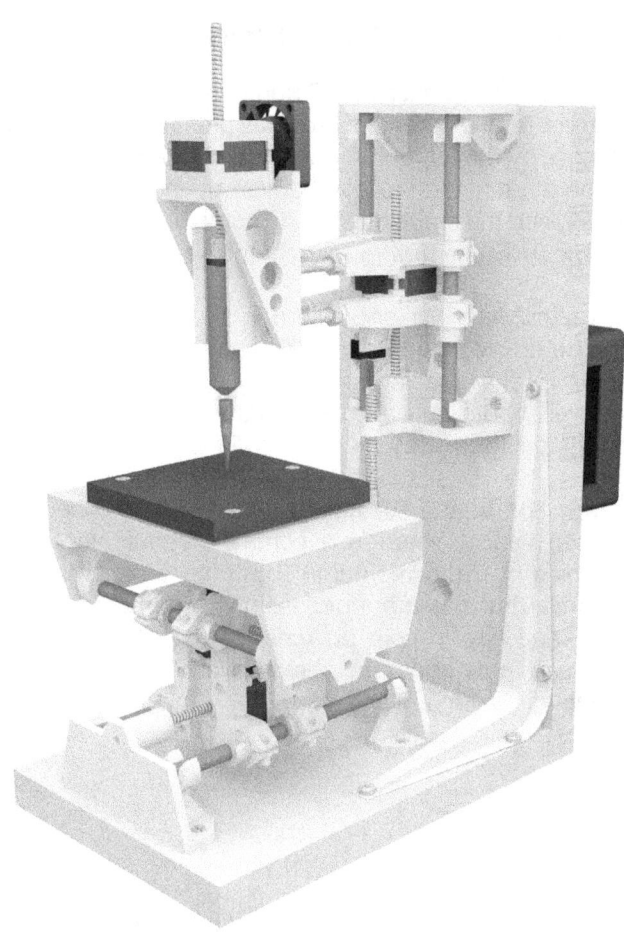

This document is intended to be used by an individual to produce an individual machine for her/himself and is not licensed for duplication or redistribution to larger groups. Additional downloads of this document may be purchased from MiniMetalMaker LLC online at www.minimetalmaker.com or through Amazon.

We appreciate your interest in our project. Thank you for your support by purchasing these instructions for yourself.

Sincerely,

David Hartkop
Chief hardware developer

I0468645

A Note From The Maker

The MMM-DIY is a fun affordable 3D printing project that lets you create a device with unique capability.

By and large, most rapid prototype machines on the market today are very similar in their use of plastic filament and belt driven mechanisms. The Mini Metal Maker is different from the rest in that it extrudes room-temperature paste media from a syringe. This capability allows for the use of metal clay as well as more generic varieties of wetted silicate type clays.

The machine motion is provided by four identical linear screw-type stepping motors. This provides unprecedented simplicity, stability and resolution. The MMM-DIY is based on the use of 3D printed components and leverages open-source software and firmware. This makes the MMM-DIY a choice platform for users who wish to modify and experiment further with the design. The Mini Metal Maker DIY and the Mini Metal Maker PRO are both products of MiniMetalMaker LLC, which successfully funded product research and development in 2013 through an indieogogo crowdfunding campaign. This publication was written as part of the fulfillment of that crowdfunding campaign.

Miniature nautilus shell pendant created in solid bronze with the Mini Metal Maker.

What is Metal Clay?

Metal clay is a relatively new material to come to the market. The first types of metal clay emerged from Japan in the 1990s in the form of silver precious metal clays. Today, metal clays are available in many various metals including gold, platinum, bronze, copper, nickel and others.

All metal clays essentially consist of metal powder held together with a water soluble binder. The material can be sculpted by hand, and is then allowed to air dry. Once dry, it is fired in a kiln like a ceramic material. The firing process burns away the binder material without disrupting the object's shape. The metal particles (typically 95%+ of the clay by weight) fuses together into a solid piece of metal.

Objects created with metal clay are metal. They are not 'like' metal, they are high purity metal. That is to say they can be polished, bent, filed and drilled. They also conduct electricity, can be annealed, oxidized, or melted down.

The Mini Metal Maker takes all of the exciting possibilities of metal clay a step further by allowing one to 3D print objects, and then render them in metal. No other low cost 3D printing system offers this capability.

What is the Mini Metal Maker?

The Mini Metal Maker is a 3D printer designed to extrude material from a syringe. It provides pressure on a syringe plunger by means of a linear drive stepper motor. Each of the three motion axis are driven by the same type of motor, and the entire machine is controlled with an open source Arduino/RAMPS microcontroller.

The Mini Metal Maker uses open source software that can be downloaded for free. This manual describes the steps to downloading and setting up one toolchain of software, but you are free to hack and use whatever programs you wish.

In addition to the DIY Mini Metal Maker planned out in this manual, there is a commercial version of the machine by the same name. The commercial machine is similar but has more professionally fabricated metal parts and has a more powerful extruder system. You can read more about the commercial Mini Metal Maker online at MiniMetalMaker.com.

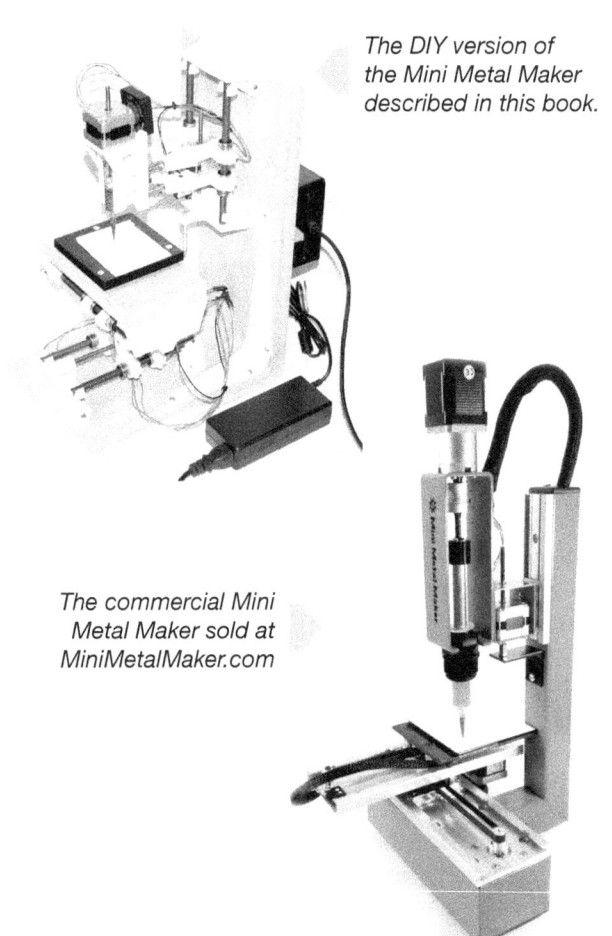

The DIY version of the Mini Metal Maker described in this book.

The commercial Mini Metal Maker sold at MiniMetalMaker.com

Mini Metal Maker pendants before and after kiln-firing. Clay may be finished and detailed with a blade before firing.

Table of Contents

1. Printed Parts

The Mini Metal Maker DIY relies heavily on a set of 3D printed parts that are assembled with off-the-shelf hardware and a few specialty motors that must be ordered to create a working machine.

1.1 Download files

Files for the Mini Metal Maker DIY are available for you to download through Github at the following address:

github.com/MiniMetalMaker/MMM-DIY

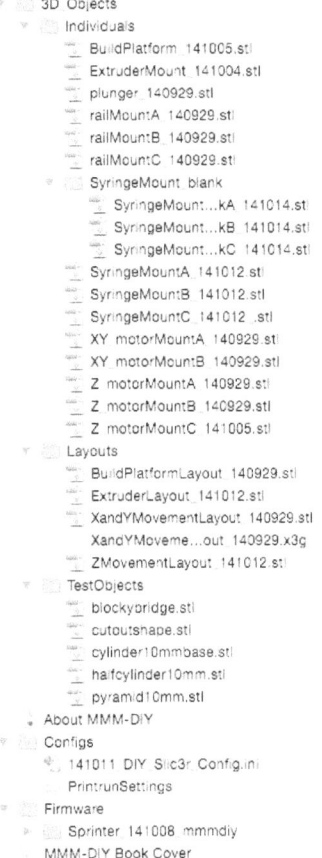

Just click to Download .Zip and you will receive a set of folders containing the project files.

The 4-digit number at the end of each file name is the creation date of that particular model in yymmdd format, depending on the MMM-DIY release you have.

The "Individuals" folder contains a complete set of individual 3D objects. The inserted number to the left of each file name denotes how many of each object should be printed in order to create the MMM-DIY.

For instance, 3 X means to print three copies of the railMountA object.

The "SyringeMount_blank" folder contains 3 object files that can be used by a skilled user of 3D modeling software to create a custom-sized syringe mount. This may be useful to users who wish to use syringes that do not fit any of the three included syringe mound sizes. If you open the three files in a 3D modeling program, it will become apparent that the shapes may be scaled and then subtracted from one another to produce a new syringe mount.

The "Layouts" folder contains .stl files of multiple objects, pre-arranged for easy 3D printing. The layouts were made to be printed on a device with a minimum print volume sizes of X = 220 mm, Y = 110 mm, Z = 55 mm. This will fit within the range of the original Makerbot Dual.

The "TestObjects" folder contains a demo object that may be sliced for printing with your finished MMM.

The "Configs" folder contains helpful settings and configuration files for different open source software used for controlling the printer. More on this later.

The "Firmware" folder contains a modified version of Sprinter firmware. This has been preconfigured to work well with the MMM-DIY. It will need to be opened in Arduino and uploaded to the controller board. More on this later too.

Please refer to figure 1.1 to see a complete visual guide to all of the objects to print.

1.2 printing on your own machine

We recommend that you print the object files using ABS but, since there are no high-temperature processes involved, the use of PLA should be just fine. In the case of PLA, there may be some issues with longevity if the device is used in a high-humidity or variable temperature environment.

The objects may be printed at the lowest resolution (fastest settings) but we recommend using at minimum infill setting of 20% for rigidity.
Be sure to turn-on your options to generate a raft and support material. The only exception may be the BuildPlatform, which can be printed directly on the hot surface for an extra-smooth top surface.

1.3 Contracted printing

If you do not have access to a 3D printer, don't worry! A first step may be to contact a public library in your area, as they may have 3D printing capability. There may also be some sort of makerspace in your area. Aside from

figure 1.1

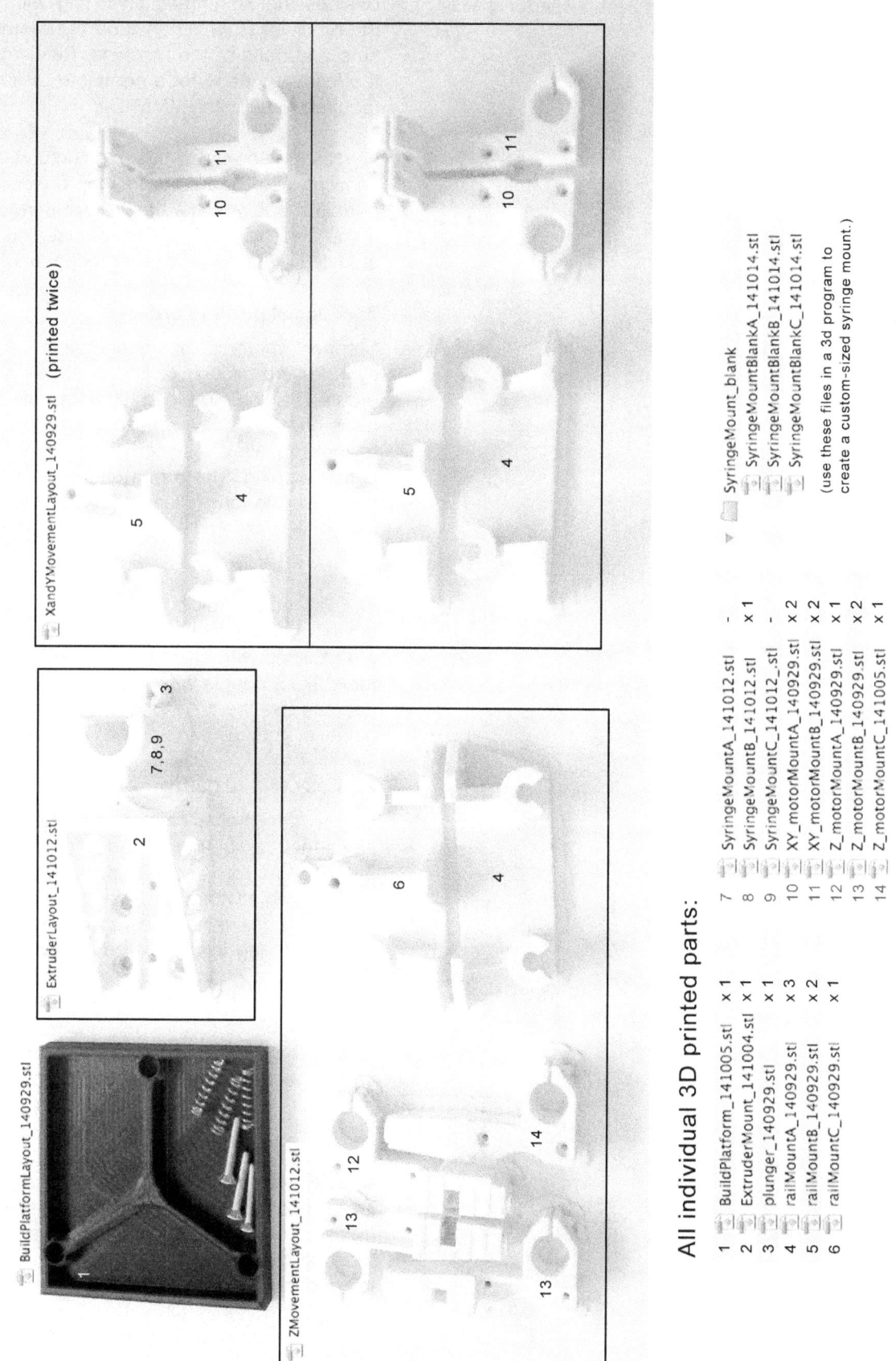

that, there are plenty of online options for having designs printed and then mailed to you. Do a Google web search for "3D Printing Services" and select a vendor that is able to quote you a price online.

2. All non-printed supplies

Besides the 3D printed files, you will need to gather the other parts, which include the motors, electronics, nuts and bolts of the machine. Refer to figure 2.1 on the following page for a complete list of all of the part required to build the MMM-DIY.

I spent way too much time making a composite picture showing all of the parts for this project all laid out on a table top. If you like that sort of thing, it is figure 2.2. If you are easily overwhelmed, just skip over it. All this stuff easily fits in one paper grocery bag. :->

2.1 Special parts to order

Stepper motors, microcontrollers and dispensing syringes are difficult to find in stores and so must be ordered online. We had little trouble locating these parts on eBay or on Amazon.com. You may also wish to look on Aliexpress.com or do a search in Google.com/shopping. The 8 mm diameter linear rails can be purchased in longer lengths and then cut with a hack saw, or else ordered pre-cut from a vendor such as Igus.com.

The list looks a bit daunting, but it pretty much boils down to tinkering with a shopping cart on eBay until you are satisfied with pricing, and then just checking out. It may take a couple weeks for all of your parts to arrive. You can use the time to organize your work space and get all of your 3D parts printed and cleaned up.

2.2 Easy to find hardware

Wood, nuts, bolts, screws and shelf brackets are all items that can be purchased at a local hardware store. Since the motors and bearings used in this design are only available with metric sized fasteners, much of the rest of the design was designed with metric sized bolts in mind. Being in the United States, however, I could not locate a metric threaded brass insert for wood. Rather than ordering it, I simply chose to use a size that was available to me, namely 8-32. Ultimately, the only screws that MUST be a metric size are the motor mount screws (see Fig 2.1 under the fasteners category.) All the others can be chosen according to availability and hole size.

I find it helpful to sort the screws for a particular project into a set of small clear plastic drawers. As an alternative, consider using labeled zip lock bags. You could go so far as to use masking tape to hang them up along a wall in your shop.

Whatever it takes to stay organized!

figure 2.1

Category	Part Name	Details	Quantity
3D Printed	Build Platform		1
3D Printed	Extruder Mount		1
3D Printed	Plunger		1
3D Printed	railMountA		3
3D Printed	railMountB		2
3D Printed	railMountC		1
3D Printed	SyringeMount_blanks	use in a 3D program to make a SyringeMount object for any syringe size	-
3D Printed	SyringeMountA	for a syringe with an O.D. of 13.7mm (B-D 5mL)	-
3D Printed	SyringeMountB	for a syringe with an O.D. of 14.2mm. (Termo 5-6mL)	1
3D Printed	SyringeMountC	for a syringe with an O.D. of 16.5mm. (CML 10mL)	-
3D Printed	XY_motorMountA		2
3D Printed	XY_motorMountB		2
3D Printed	Z_motorMountA		1
3D Printed	Z_motorMountB		2
3D Printed	Z_motorMountC		1
clay related	metal clay, powder form	metal clay of your choice in powder form	1
clay related	mixing canister	plastic film canister for mixing metal clay with water	10
clay related	mixing sticks	round wooden chopsticks	10
clay related	plastic extrusion tips	22 gage blue, can be ordered from CML Supply, sold in packs of 50	1
clay related	syringes	recommend starting with Termo 5-6mL (O.D. of 14.2mm)	10
electronics	electronics box	5cm x 10cm x 15 cm plastic electronics box	1
electronics	end-stop switches	RU SS-5F or similar microswitches	3
electronics	fan	4cm x 4cm x 1cm 12VDC PC fan	1
electronics	Hook-up wire	22 gage stranded wire, 25 ft spools in Red, Blue, Green, Yellow, Black	1
electronics	linear-movement motors	Wantai 39BYGL215A Nema 16 linear stepper	4
electronics	microcontroller	Arduino Mega 2560	1
electronics	motor controller board	RAMPS shield for Arduino	1
electronics	motor-to-board connector	Dupont wire jumper pin header connecter housing 1x4 Male / Female	4
electronics	pin header crimp inserts	Dupont female servo crimp wire jumper pin header connector	36
electronics	power jack	12VDC threaded power jack	1
electronics	power supply	12VDC 6A LCD power supply	1
electronics	power switch	SPST toggle switch	1
electronics	stepper motor drivers	Allegro A4988 DMOS motor drivers	4
electronics	switches-to-board connector	Dupont wire jumper pin header connecter housing 1x2 Male / Female	4
fasteners	bearings mount screws	M2 x 8mm pan head	12
fasteners	extruder extending nuts	M6 nuts	12
fasteners	extruder extending screws	M6 x 6cm pan head	4
fasteners	leveling screw inserts	8-32 brass threaded inserts for wood	3
fasteners	leveling screws	#8-32 x 1in long	3
fasteners	motor mount screws (long)	M3 x 30mm pan head	15
fasteners	motor mount screws (short)	M3 x 8mm pan head	16
fasteners	PCB mount washers	M3 split lock washers	3
fasteners	PCB mount nuts	M3 nuts	3
fasteners	PCB mount spacers	M3 x 8mm long nylon female threaded spacer	3
fasteners	plunger set screw	M3 x 3mm set screw	1
fasteners	switch / fan mount screws	M2 x 12mm pan head	8
fasteners	syringe mount nuts	M4 nuts	2
fasteners	syringe mount screws	M4 x 20mm pan head	2
fasteners	syringe mount washers	M4 flat washers	2
fasteners	wood screws (pan head)	16mm long pan head wood screws	12
fasteners	wood screws (tapered)	16mm long taper head wood screws	17
fasteners	Z-endstop adjustment screw	M6 x 5cm (1/4-20 x 2.5in) threaded rod	1
hardware	90 degree support brackets	7.5in x 5.24in L shelf bracket	2
hardware	adhesive rubber feet	self adhesive rubber feet	4
hardware	leveling springs	O.D.=1/4in, L=1in, wire O.D.=1/32in	3
hardware	linear bearings	LM8UU linear bearings for 8mm round rails	12
hardware	linear rails	8mm O.D. Smooth round rail, 16cm long	6
wood	back board	18cm x 35cm x 2cm thick	1
wood	base board	18cm x 24cm x 2cm thick	1
wood	build platform base	18cm x 10cm x 2cm thick	1

figure 2.2

12VDC 6A LCD power supply

12VDC power jack

Switch

3 X M3 x 8mm threaded nylon PCB spacer

1 X 3m x 3mm set screw

2 X M4 x 20mm pan head + 2 nuts & 2 washers

15 X M3 x 30mm pan head + 3 nuts

16 X M3 0.5 x 8mm pan head

8 X M2 x 12mm pan head

12 X M2 x 8mm pan head

17 X 16mm taper wood screw

12 X 16mm pan head wood screw

4 X M6 x 6cm pan + 12 nuts

1X M6 x 5cm Threaded rod

5 x 10 x 15 cm electronics box

18 x 10 cm board, 2 cm thick

3 x 8-32 brass threaded inserts

wires with 2 & 4 pin wire jumper pin header connectors (see page bottom)

12 x LM8UU linear bearings

Allegro A4988 DMOS motor drivers

Arduino Mega 2560 & RAMPS shield

18 x 24 cm board, 2 cm thick

4 X adhesive rubber feet

36 X Female Dupont and Servo Crimp Wire Jumper Pin Header Connector

4cm sq. 12VDC PC fan

3 x RU SS-5-F Microswitches

4 x stepper motor Wantai 39BYGL215A

6 x 8mm dia rail, 16 cm long

18 x 35 cm board, 2 cm thick

90° support brackets

4 X Dupont Wire Jumper Pin Header Connector Housing - 1x2 - Male / Female

4 X Dupont Wire Jumper Pin Header Connector Housing - 1x4 - Male / Female

All 3D printed parts

3 x leveling screws: taper #8-32 x 1"
3 x springs: O.D.=1/4", L=1", wire O.D.=1/32"

Thermo 6mL syringe

22 gage dispensing tips

3. Tools and Workspace

3.1 Tools needed

Measurement
  Tape measure
  Ruler
  Carpenter's square
  Mechanical pencil
  Black marker

Wood Working
  Circular saw or hand saw
  Sand paper
  Drill
  Box of assorted drill bits
  Phillips head screwdriver
  Needle nose pliers

Metal Working
  Hack saw, fine tooth
  Metal file, flat, fine tooth.
  Vise or C-clamp

Plastic clean-up
  Small blade utility knife (Exacto etc.)
  Butane lighter (optional)

Electronics
  Soldering iron
  Rosin core solder
  Vinyl electrical tape
  Wire strippers
  Volt/Ohm meter (optional)
  Shrink-Wrap tubing (1/8" or 3 mm)

Computer Equipment
  PC or Mac computer
  Internet connection
  Permissions to install software
  USB printer cable

Misc.
  Bright adjustable desk lamp
  Ability to play music for hours
  Coffee :-)

3.2 Workspace

I am always surprised how little space is actually needed to accomplish DIY projects. Most spaces, be they bedrooms, home offices, a covered porch or garage provide ample space. That said, more space is better!

This project is part construction, part model-making, part electronics, and part computer science. I have tried to break up the instructions into one sort of task at a time. Hopefully this will allow you reconfigure your space efficiently, and not have to keep switching between, for instance, wood-shop and computer lab.

The MMM is not a very big machine, but there are so many parts that it may become unmanageable unless you have at least a full table top to lay out all of the parts and the instructions.

Be sure that you have ample ventilation when using a soldering iron or a butane lighter to smooth plastic or heat shrink wrap tubing.

Light is extremely important, especially when dealing with small screws or trying to line up parts to bolt down with any accuracy. Having a well lit work space is also safer in terms of tool use, and lends to a better state of mind.

In my opinion, there is nothing more fun or rewarding than a good do-it-yourself maker project - especially with good music playing, ample caffeination, and the world to yourself in the hours after 1 a.m.!

(Depending on your living situation, you may wish to do the first step - which involves wood working power tools - during the daylight hours.)

4. Machine Base

The base structure of the MMM-DIY is made from two pieces of wood held together by a pair of shelf brackets.

4.1 Cut back board, base board

Use a circular saw, table saw, or a hand saw to cut a 2 cm thick plank of pine or ash into two rectangular boards as shown below in figure 4.2.

4.2 Drill wire holes in the back board

Use a carpenter's square and pencil to measure and mark the positions of the wire holes. (See figure 4.1 at right.) Use a drill with a large circular bit or a flat wood bit to drill two holes of approximately. 1.5 cm diameter in positions marked.

4.3 Mount shelf brackets

It may be helpful to mark through the holes of the brackets with a pencil and pre-drill the holes with a small sized bit. *NOTE: The back board abuts the back end of the base board - *it does not go on top of it!*

figure 4.1

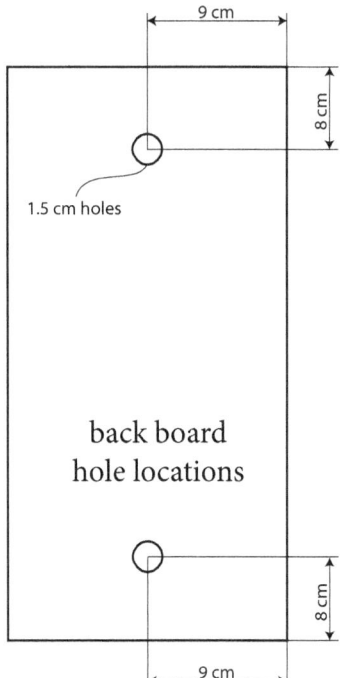

1.5 cm holes

9 cm

8 cm

8 cm

9 cm

back board
hole locations

figure 4.2

18 cm

35 cm

pan head
wood screws

shelf
brackets

back board

base board

18 cm

24 cm

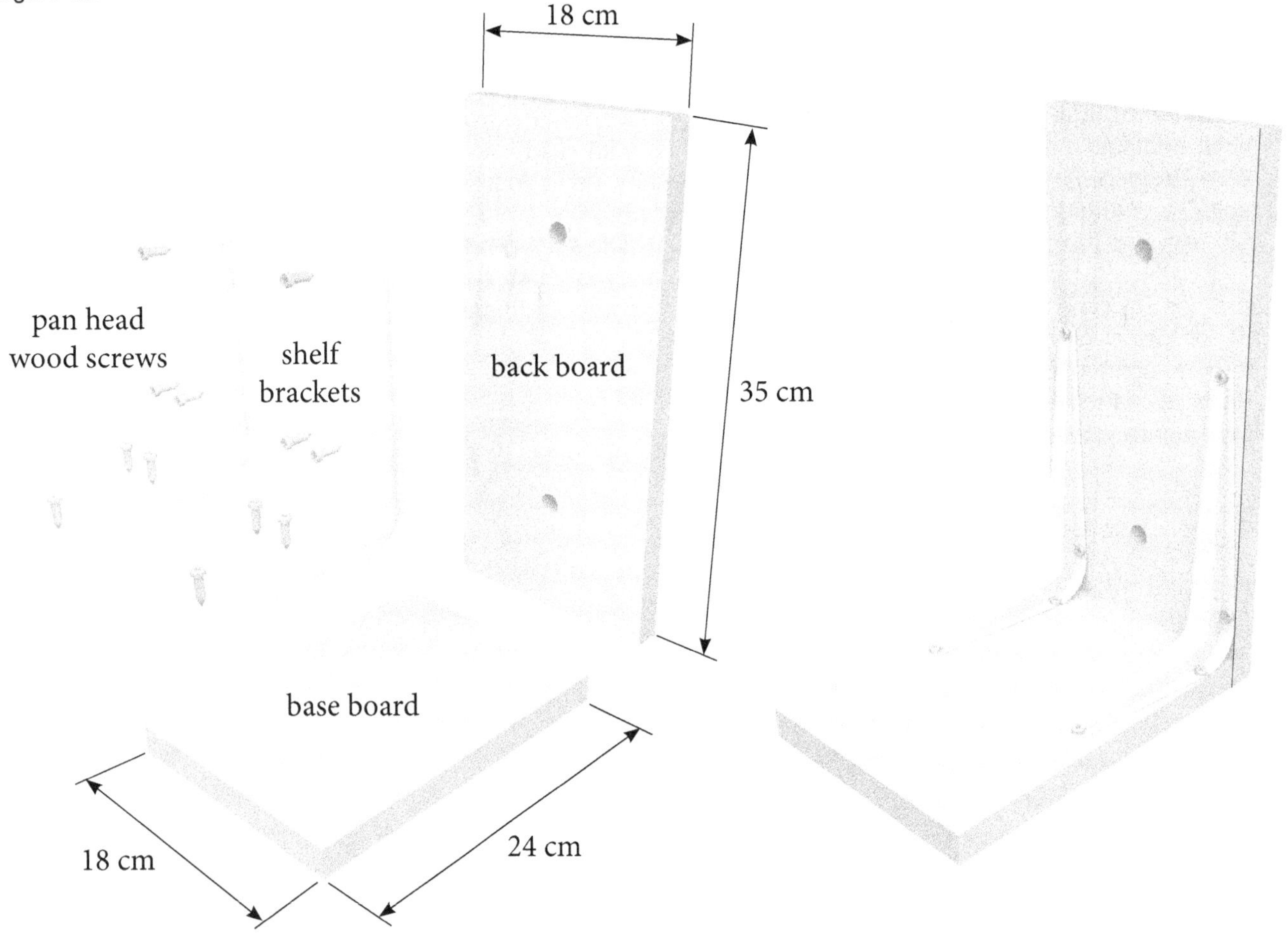

5. X-Y Carriage

The X-Y Carriage provides the movement of the build platform in the X and Y directions.

5.1 Assembly

Parts you need:
    2 x Wantai 39BYGL215A stepper motors
    4 x 8 mm dia x 16 cm long linear rails
    2 x XY_motorMountA (3D printed)
    2 x XY_motorMountB (3D printed)
    8 x LM8UU linear ball bearings

Screws you need:
    8 x M3x30mm pan head
    8 x M3x8mm pan head
    8 x M2x12mm pan head
    16 x M3 split lock washers

Mount the 3D printed XY_motorMount parts on the four corners of the stepper motor frames using the M3 pan head machine screws.

Note, you will have to remove the set of four long screws from the motor frame and replace them with the M3x30mm screws. Care should be taken in the order you do this because, if all four bolts are removed from a motor at once, the motor tends to spring apart. Do one corner at a time.

Split ring lock washers should be used as spacers on the four corners that are indented. The spacers will allow tightening of the bolts without cracking the 3D printed parts.

The linear bearings can be inserted by hand as shown in figure 5.1B. They simply slip into place and click. Test the alignment by slipping a linear rail through each pair of bearings as you insert them.

Tighten the clamps around the bearing with the M2x12 mm screws. The screws do not necessarily need to be very tight.

figure 5.1A

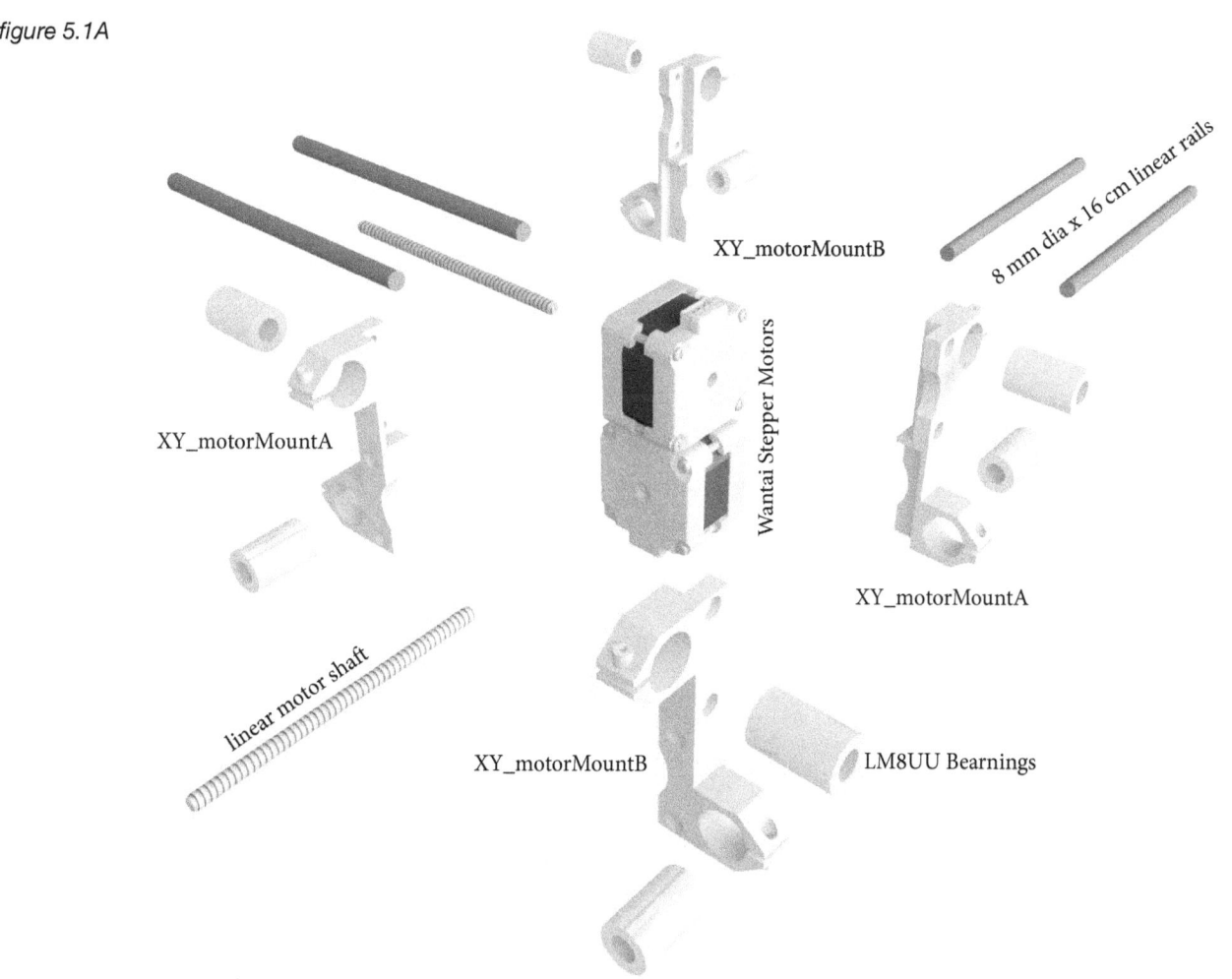

XY_motorMountB

8 mm dia x 16 cm linear rails

XY_motorMountA

Wantai Stepper Motors

XY_motorMountA

linear motor shaft

XY_motorMountB

LM8UU Bearnings

figure 5.1B

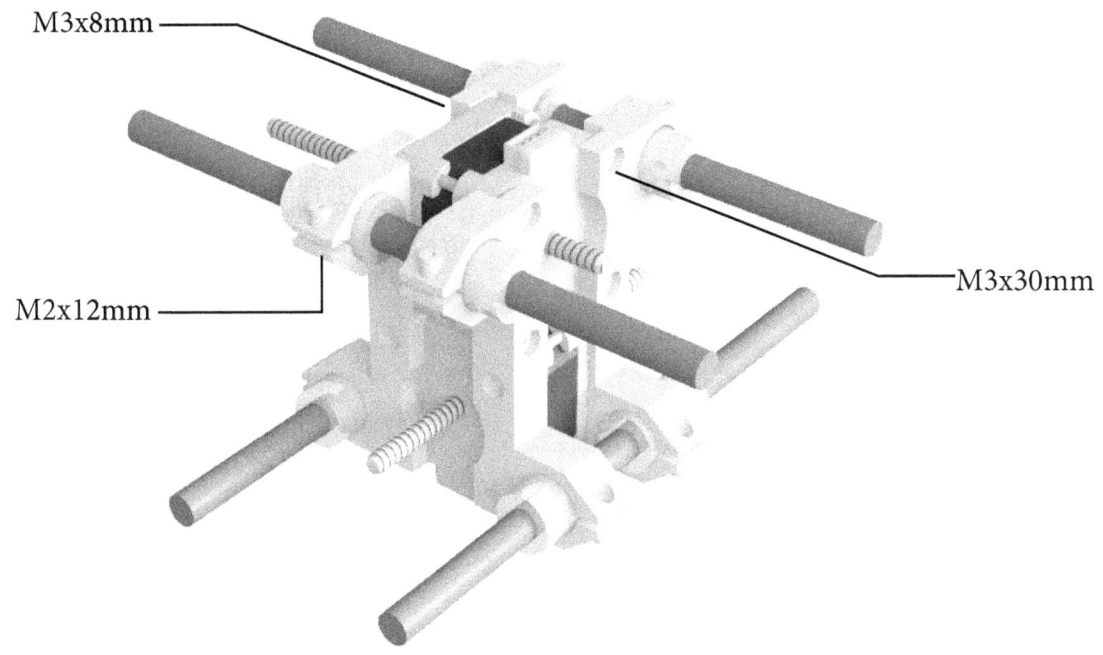

M3x8mm

M2x12mm

M3x30mm

figure 5.2A

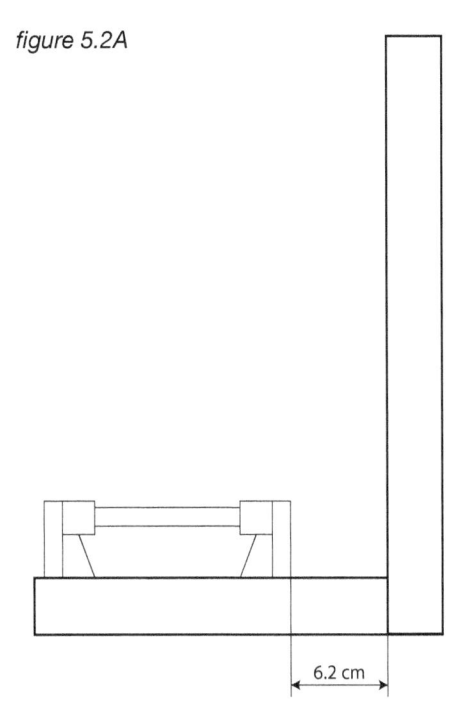

6.2 cm

5.2 XY Carriage Mounting

Parts you need:
>   2 x railMountA (3D printed)
>   2 x railMountB (3D printed)
>   2 x end-stop microswitches
>   1 x wooden base
>   Wood to make build platform base
>   Assembled XY carriage block & rails

Screws you need:
>   10 x tapered wood screws
>   4 x M2x12 mm pan head

Cut a rectangular piece of wood to 18 cm x 10 cm. This will serve as the build platform base.

Bolt the endstop microswitches into place on the railMountB pieces with the M2x8 mm pan head screws. You may wish to solder a pair of 18" (45cm) long wires to each switch before mounting. It is difficult to solder once the switches are mounted.

Pop the railMountA and railMountB pieces onto the ends of the linear rails. The railMountA piece should be located at the left of the carriage system on the top, and should be the side closest to the back board at the bottom (figure 5.2B) You can now place the unit onto the base in order to mark through the holes with a mechanical pencil. Space the carriage system on the machine base so that the back edge of the railMount is 6.2cm from the back board (see figure 5.2A)

Mark through the holes onto the build platform base. Use a drill with a small bit to pre-drill the marked hole positions. Remove the railMounts from the linear rails. Use a screwdriver to attach the railMounts to the machine base and the build platform base with tapered wood screws.

Screw the threaded drive shaft half-way through each of the stepper motors so that the threaded end points into the railMountB piece for the top and bottom respectively.

Firmly press the XY carriage assembly's linear rails into the mounted railMount pieces. The carriage assembly will pop firmly into place. Finally, use needle nose pliers to grip the drive shafts of the motors by the un-threaded tip. Press and turn the shaft into the socket of each railMountB piece until the shaft is screwed in 2 or 4 turns. (See figure 5.2C) Once both shafts are screwed into the railMountB pieces, the XY carriage system fully assembled.

figure 5.2B

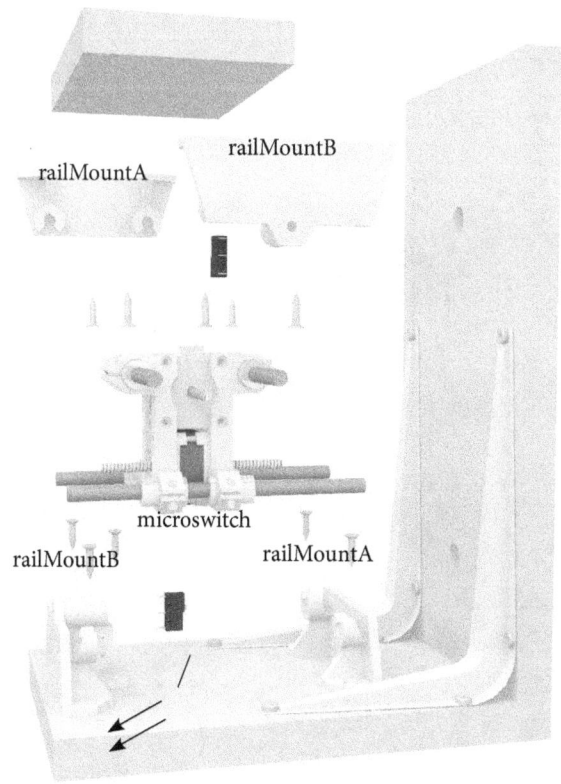

figure 5.2C

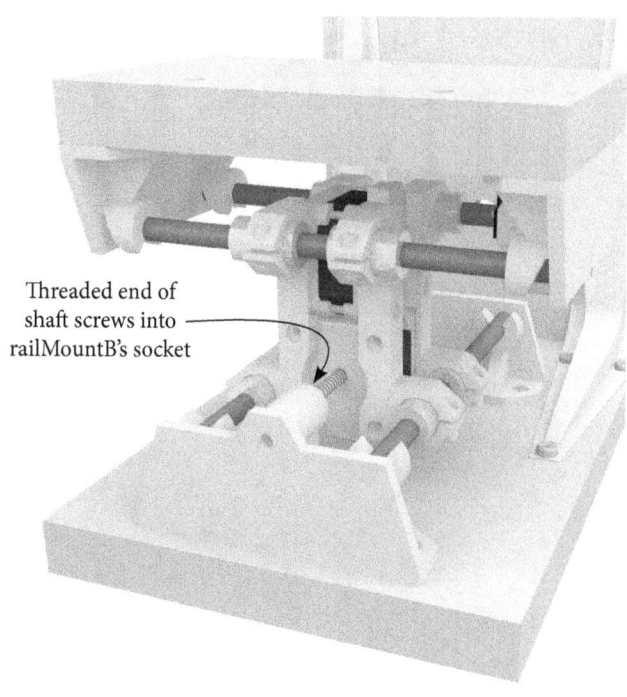

6. The Build Platform

The build platform is the square state onto which your 3D objects will be printed. It is equipped with three screws for leveling.

6.1 Assembly

Parts you need:
    1 x Build Platform (3D printed)
    3 x leveling springs
    3 x #8-32 threaded brass inserts

Screws you need:
    3 x #8-32 tapered screws, 1in long

Place the build platform piece centered on the wooden build platform base with two screw holes facing away from the back board, as pictured in figure 6.1. Mark through the holes with a mechanical pencil.

Drill holes all the way through the wooden build platform base with a bit of appropriate size for the threaded inserts you have selected. Note: you may wish to remove the wooden build platform base from the XY carriage assembly in order to place under a drill press for perfectly vertical bore holes.

Once the holes are drilled, insert threaded inserts by hand with a screwdriver such that they are near the very bottom of the holes. This will leave compression space for the leveling springs within the wooden bore holes.

6.2 Mounting Build Platform

Place the tapered screws through the top (smooth side) of the 3D printed build platform. Position a leveling spring over each screw end, holding in place with fingers. Align the screws into the holes in the wooden build platform base. While pushing the build platform down against the springs force, use a screwdriver to tighten each screw into its threaded insert within the wooden build platform base. (see figure 6.2.)

The platform should be firmly seated, but have room to move down against the force of springs if each of its corners are pressed. You do not need to screw in each screw all the way, as long as it is engaged with the threaded insert. The platform will be leveled in a later step.

figure 6.1A   *figure 6.2*

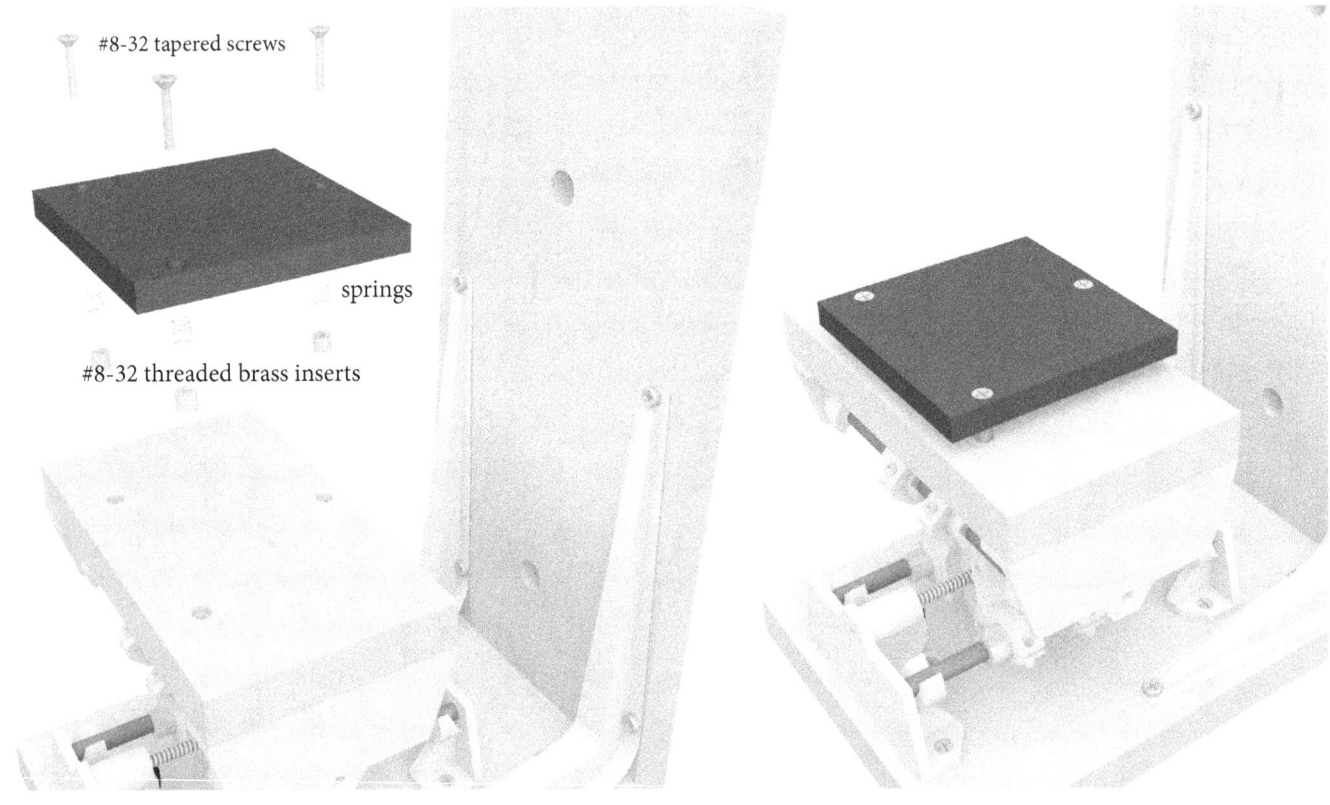

#8-32 tapered screws

springs

#8-32 threaded brass inserts

7. Z Carriage

The Z Carriage provide movement to the extrusion head in the Z axis.

7.1 Assembly

Parts you need:

- 1 x ZmotorMountA (3D printed)
- 2 x ZmotorMountB (3D printed)
- 1 x ZmotorMountC (3D printed)
- 4 x linear bearings
- 1 x linear stepper motor
- 1 x endstop microswitch
- 8 x M3 split ring washers.

Screws you need:

- 4 x M3 x 30 mm pan head screws
- 4 x b x 8 mm pan head screws
- 6 x M2 x 12 mm pan head screws

Bolt the ZmotorMount pieces on each of the four corners of the linear stepper motor's body as pictured in figure 7.1A using the M3 screws. Note: You will need to remove the existing M3 screws from the motor. If you remove all four at once, the motor will spring apart, so be careful.

Insert two split ring washers between the ZmotorMount and the motor to act as spacers on the end of the motor that has indented screw holes. This will help to prevent cracking of the motorMounts if the screws are overtightened.

Bolt the microswitch into place on the tab extending from the bottom of ZmotorMountC using the M2 x 12 mm screws. You may wish to solder a pair of 18" (45 cm) long wires to the switch before mounting. It is difficult to solder once the switch is mounted.

Insert the linear bearing into the bearing clamps as shown in figure 7.1B. Insert a pair of linear rails as shown. Use the M2 x 12 mm screws to tighten the clamps around the bearings as needed to firmly seat them.

figure 7.1A

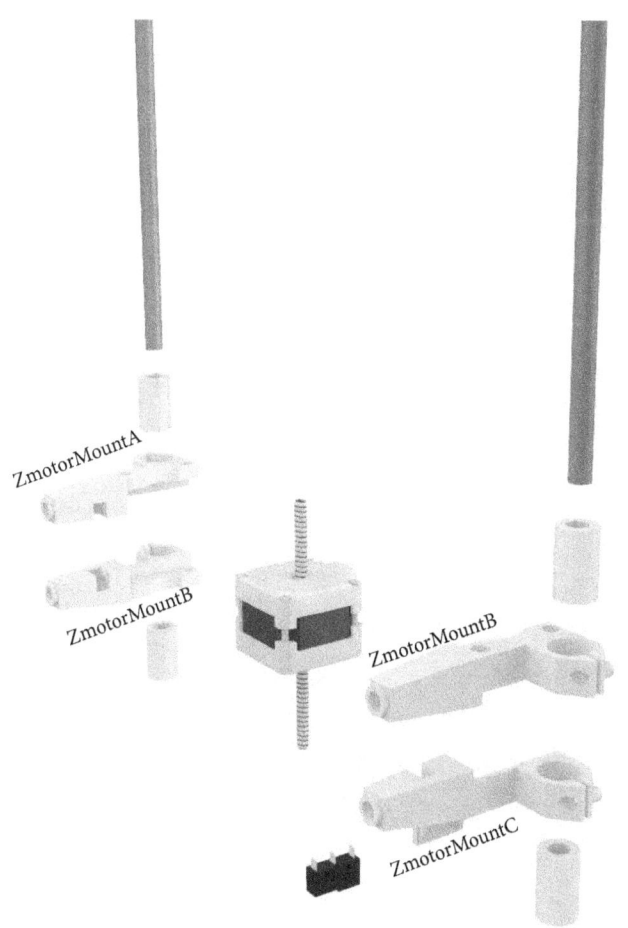

ZmotorMountA

ZmotorMountB

ZmotorMountB

ZmotorMountC

figure 7.1B

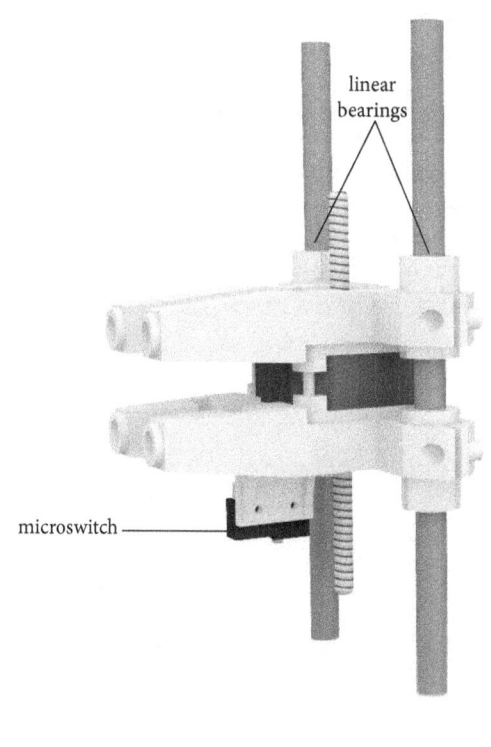

linear bearings

microswitch

7.2 Mounting the Z Carriage

figure 7.2A

Parts you need:
  1 x railMountA (3D printed)
  1 x railMountC (3D printed)
  Assembled Z Carriage block & linear rails

Screws you need:
  5 x taper head wood screws
  Z-endstop adjustment screw
    (5cm long piece of M6 or 1/4-20 threaded rod)

Pop the linear rails into the railMounts A & C, with A at the top and C at the bottom, as shown in figure 7.2B. Position the assembly on the back board, centered horizontally. Insure that the bottom most edge of railMountC is 16.2cm up from the base board (see figure 7.2A.) Mark through the holes with a pencil.

Pre-drill the holes with a small bit. Remove the railMounts from the linear rails and mount them by screwing wood screws into their holes by hand.

Insert the threaded shaft into the linear stepping motor and center half-way through the motor. Firmly press the linear rails into the railMounts until they pop into place.

Use a pair of needle nose pliers to grip the un-threaded end of the threaded shaft (which should be facing up.) Press downward and turn the shaft by hand in order to screw the threaded end into the socket in railMountC 3 to 4 turns. See figure 7.2C.

Finally, screw the Z-Endstop adjustment screw into the small hole that extends from railMountC. Note: The end of this screw should line up to depress the microswitch on the Z carriage. Adjusting this screw up and down by hand provides fine control of the Z axis' zero position. See figure 7.C.

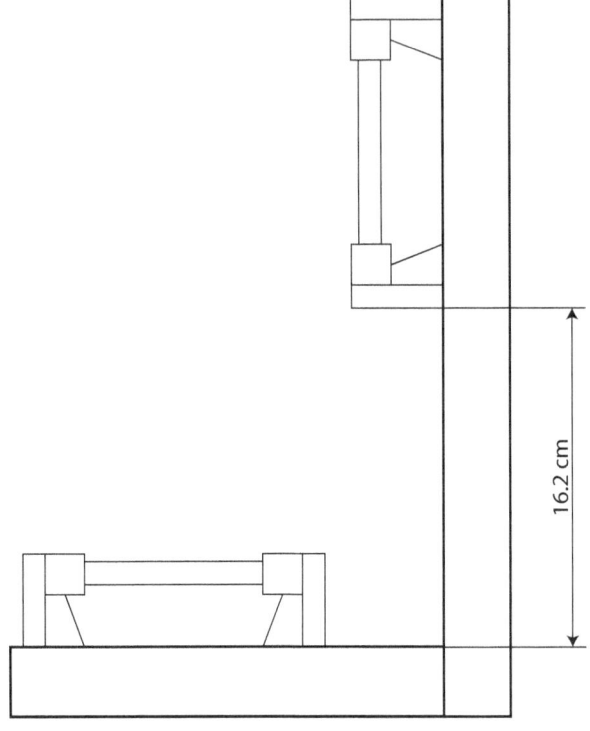

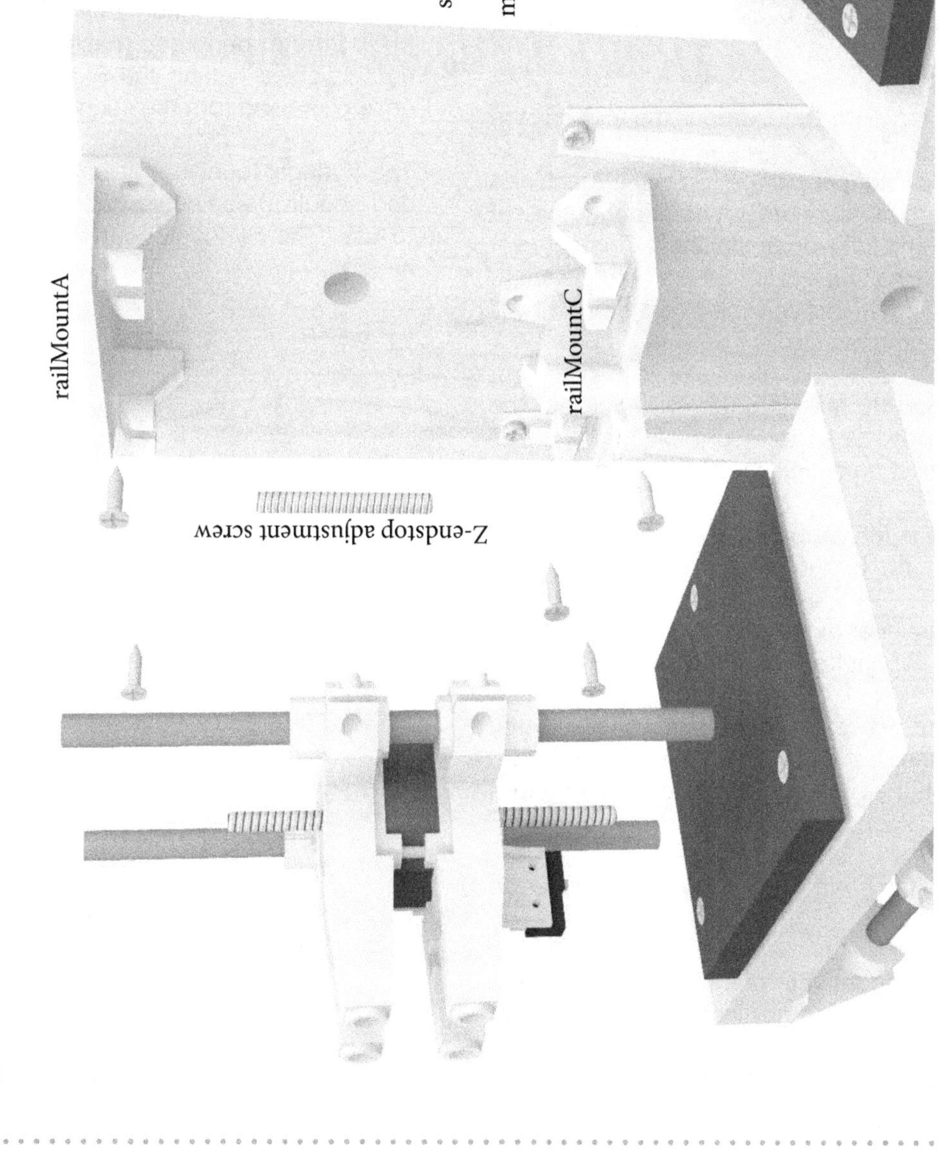

figure 7.2C

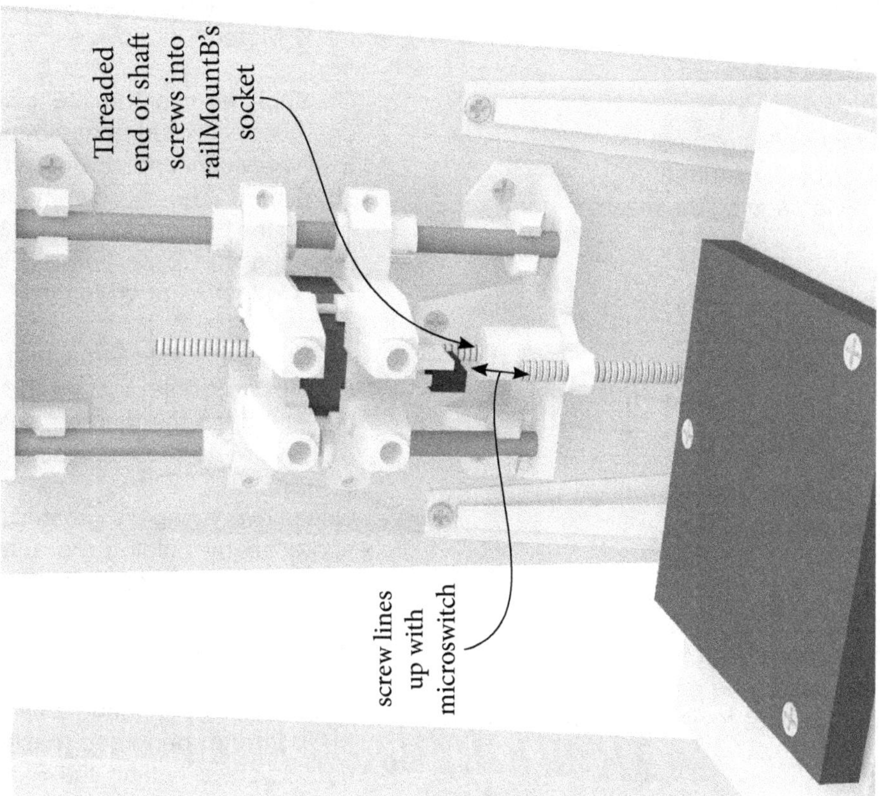

Threaded end of shaft screws into railMountB's socket

screw lines up with microswitch

railMountA

railMountC

Z-endstop adjustment screw

figure 7.2B

8. Extruder Module

8.1 Assembly & Mounting

Parts you need:
  1 x ExtruderMount (3D printed)
  1 x SyringeMount (3D printed)
  Assembled Z Carriage on the machine body

Screws you need:
  4 x M3x8mm pan head screws
  2 x M4x20mm pan head screws
  2 x M4 nuts + washers
  4 x M6x6cm pan head screws (1/4-20 x 2.5")
  12 x M6 nuts (or 1/4-20)
  8 x M6 flat washers
  8 x M6 split ring lock washers
  2 x M2x12mm pan head screws
  1 x M2 set screw

Bolt the stepper motor body onto the extruder mount with 4 x M3x8mm screws. You may use M3 size split ring washers if you wish. The connection port should face to either side, and not toward the fan. See figure 8.1A.

Bolt the 12VDC fan onto the extruder mount near the motor with two M2x12mm screws.
Place the small cylindrical plunger onto the machined end of the linear motor's threaded shaft so that the set screw hole faces the flat side of the shaft. Tighten a set screw into the hole so that it is firmly held onto the shaft. Note: overtightening the set screw is bad because it will strip the plastic of the plunger.

Insert each of four M6 nuts into the slots cut into the inner corners of the Z_MotorMount pieces in the Z carriage. The nuts should just fit into place.

Slide each of four M6x6cm screws through the four mounting holes in the ExtruderMount.

Place one flat washer and one split ring washer on each screw, slid up to the flat bottom of the extruder mount.

Place two M6 nuts on each of the screws. Turn them until they are finger-tight on the back of the Extruder mount, then back them off so that the screws can turn freely.

Place one flat split ring washer then one flat washer on each of the screws.

Hold the extruder mount with one hand, and align the four M6 screw ends so that they insert into the holes of the Z_MotorMount pieces in the Z carriage. Turn each of the M6 screws with a screwdriver while lightly pushing.

This will thread each screw into the nuts in the slots of the Z_MotorMount pieces.

Continue turning the M6 screws with a screwdriver until the measured space between the flat back of the ExtruderMount and the tips of the Z_MotorMount pieces is 25 mm. (The measurement should be made when pulling the ExtruderMount away from the Z_MotorMount pieces, as the screw heads will be seated firmly into the holes in the ExtruderMount when everything is tight.)

When you are satisfied with the spacing, finger tighten each of four nuts against the back of ExtruderMount. Finger tighten the four remaining nuts against the ends of the Z_MotorMount pieces. See figure 8.1B.

Once everything is seated, nuts may be further tightened by holding the screws' rotations stationary with a screwdriver and by turning each nut with a pair of needle nose pliers. Be careful not to crack the plastic by over-tightening.

Attach the SyringeMount to the ExtruderMount with two M4 x 20 mm pan head machine screws. Hold the M4 nuts in place behind the Extruder Mount with needle nose pliers and turn the screws with a screwdriver.

The SyringeMount will fit one specific size of syringe, and should be selected based on the type you intend to use. The SyringeMount can always be replaced if needed.

figure 8.1B

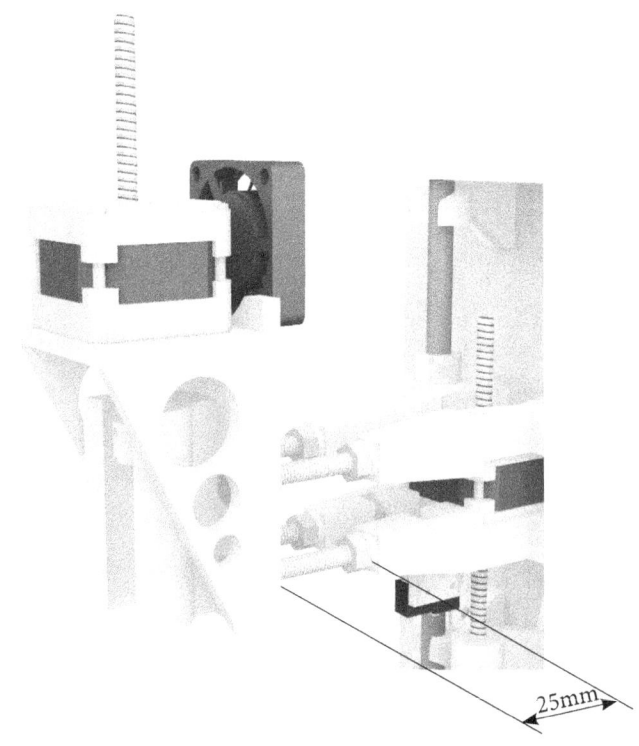

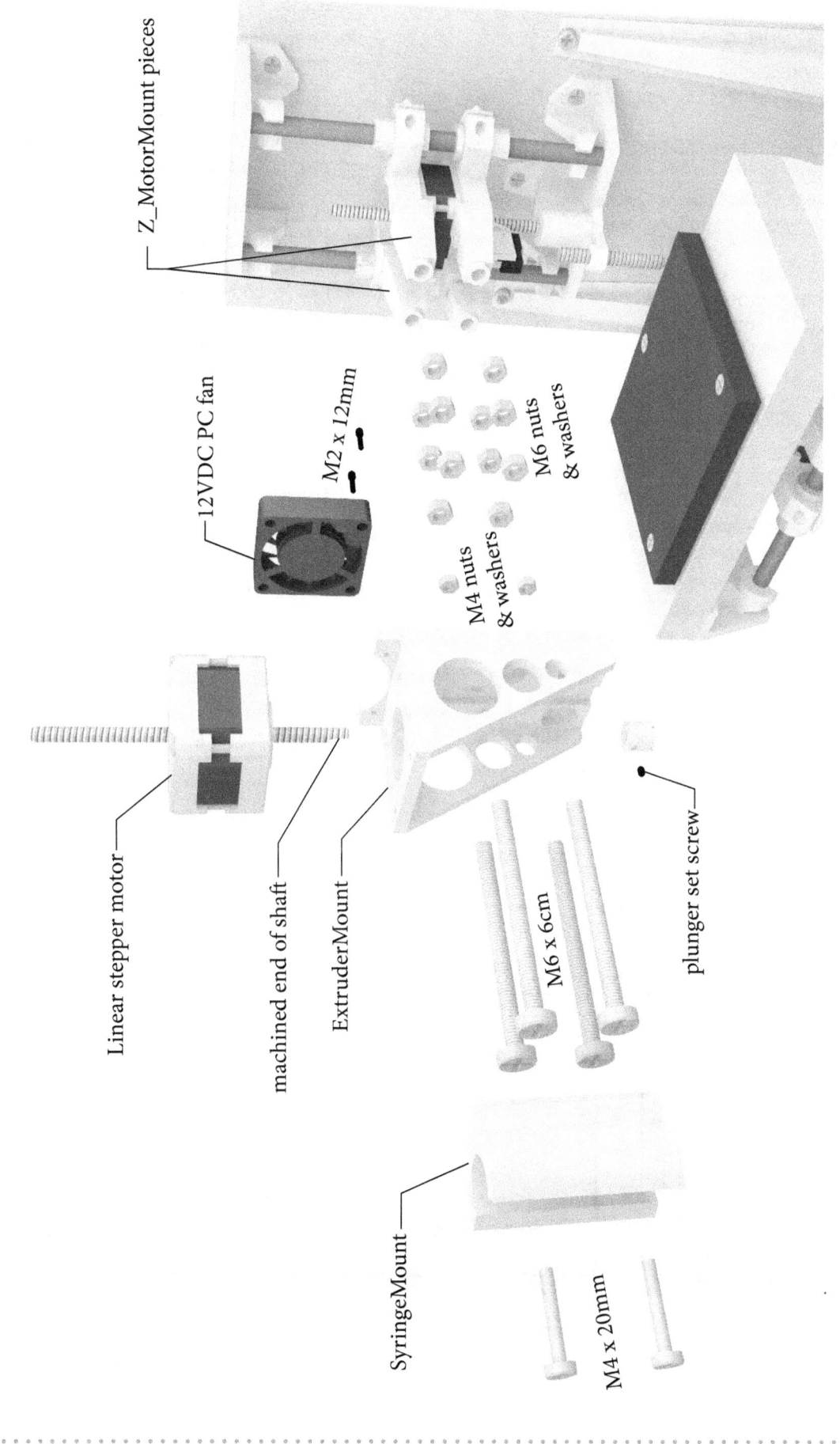

figure 8.1A

Z_MotorMount pieces

12VDC PC fan

M2 x 12mm

M6 nuts & washers

M4 nuts & washers

Linear stepper motor

machined end of shaft

ExtruderMount

M6 x 6cm

plunger set screw

SyringeMount

M4 x 20mm

9. Electronics Box

9.1 Cutting & Drilling

Parts you need:
    1 x plastic electronics box 15 cm x 10 cm x 5 cm
    deep
    1 x Arduino Mega 2560 & RAMPS microcontroller
    1 x toggle switch
    1 x female 12VDC power jack
    3 x M3 nylon female threaded PCB spacers

Screws you need:
    2 x tapered wood screws
    4 x M3x30 mm pan head screws
    3 x M3 nuts & washers

The plastic electronics box will contain the electronic controls for your Mini Metal Maker DIY. Under normal running conditions, the board will not become sufficiently hot as to require its own fan.

Various slots and holes will be cut in the box to provide access for wires. These openings also provide ventilation.

If you find that the board is running warmer than you are comfortable with, you may choose not to close the box lid. Alternatively, you may choose to install an additional PC fan.

Use a drill to drill holes in the top and bottom of the box as shown in figure 9.1. The exact hole sizes will depend on your choice of switch and 12VDC power jack.

Cut a slot in the end of the box as shown in figure 9.1. This will provide access to the USB connector and the auxiliary connections of the microcontroller. It may be easiest to first drill each of the four corners of the slot with a small drill bit, and then connect the holes by slicing through the plastic with the heated tip of a utility knife.

Note: If you are using a hot knife to cut plastic, be sure to work in a well ventilated area.

9.2 Power port & switch

Install the power switch and the power port through the top left and bottom left holes, respectively. Refer to figure 9.1.

figure 9.1   5cm x 10 cm x 15 cm electronics box

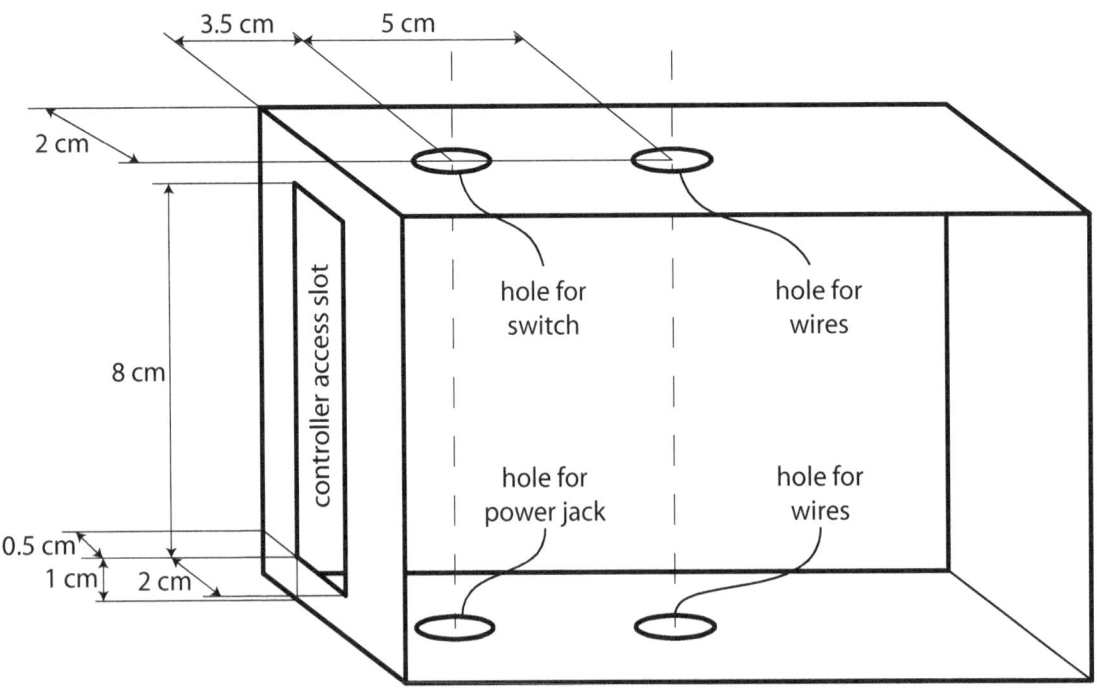

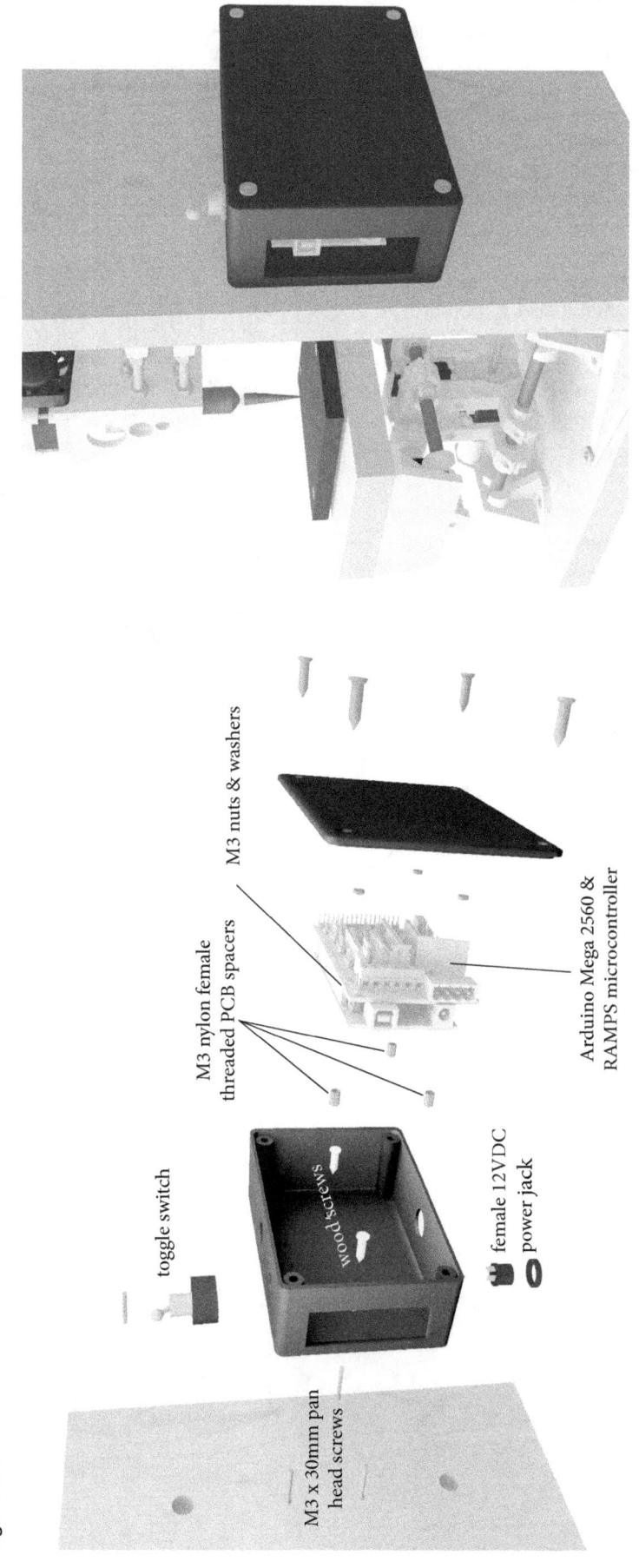

figure 9.2A

M3 nuts & washers

M3 nylon female
threaded PCB spacers

Arduino Mega 2560 &
RAMPS microcontroller

toggle switch

wood screws

female 12VDC
power jack

M3 x 30mm pan
head screws

9.3 Arduino/Ramps Microcontroller

Next, mount the Arduino/RAMPS microcontroller block into the bottom of the box.

Hold the Arduino/RAMPS microcontroller block so that the USB connector faces toward the open controller access slot. Mark through the three open PCB holes by turning the end of a small phillips head screwdriver against the plastic.

Use a small sized drill bit (M3) to drill a clean hole through each of these three marks. Insert three screws size M3x30mm through the back of the box so they face inward into the box. Tighten each of three M3 nylon spacers down onto the screws inside the box. This provides a secure set of standoffs for the controller board. See figure 9.2A.

9.4 Mounting

Mount the electronics box onto the back of the wooden upright Back Board of the machine base. Center the box vertically between the two holes through the wooden Back Board, and attach with two tapered wood screws.

You should pre-drill two holes through the box and taper the holes with a large bit size so that the screw heads are flush into the bottom of the box.

Finally, push the Arduino/RAMPS board onto the three mounting posts within the box & secure with three M3 nuts. Figure 9.2B shows how the finished box will look, once the components are wired.

figure 9.2B

Wires to
Fan, E-Motor, Z-Motor,
& Z-endstop switch

Power Switch

Arduino Mega 2560
& RAMPS shield
4 x Allegro A4988
DMOS drivers

12 VDC power jack

Wires to
X-Motor, Y-Motor,
X-endstop switch, Y-endstop switch

figure 10

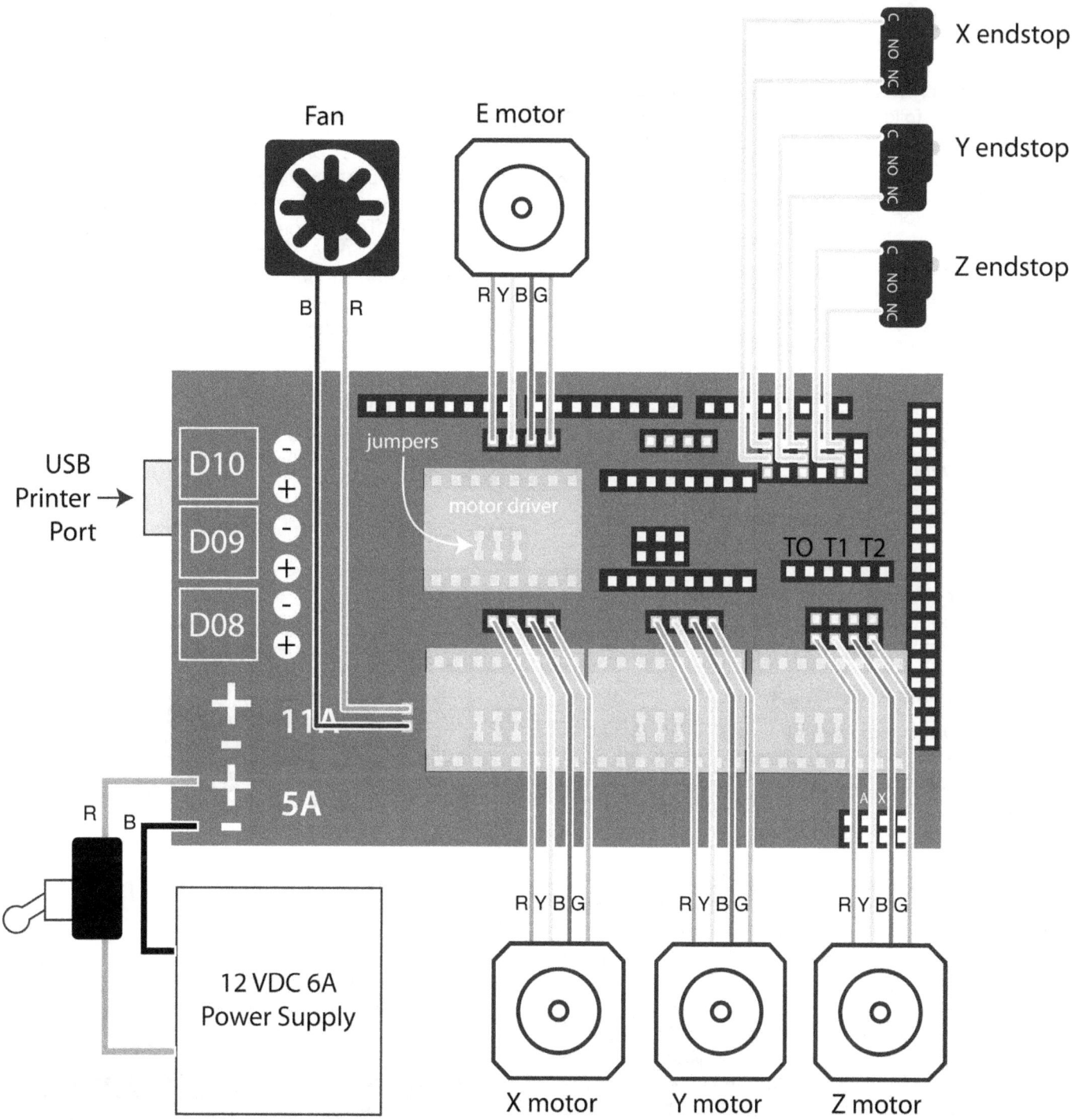

10. Wiring

The Mini Metal Maker DIY is actually less complex than most 3D printers. The system does not require a heated platforms, heated print heads, nor does it require any temperature sensors.

A complete schematic of the MMM-DIY is shown in figure 10. The following steps are intended to assist you in wiring your device to match this schematic.

Tools you need:
    soldering iron
    solder
    small needle nose pliers or clamps
    butane lighter (for shrink wrap tubing)
    wire cutters
    wire strippers
Parts you need:
    red wire
    black wire
    blue wire
    yellow wire
    green wire
    1/8in or 3mm shrink wrap tubing
    4 x Allegro A4988 DMOS stepper drivers
    12 x PCB jumpers (included with step driver)
    4 x Dupont Wire Jumper Pin Header 1x4
    4 x Dupont Wire Jumper Pin Header 1 x 2
    36 x Female crimp wire jumper pins

10.1 Hook up the Power Supply

The MMM-DIY is powered by a 12VDC 6A regulated power supply of the sort commonly used for LCD computer monitors. The jack from this power supply should be able to plug into the female power jack in the bottom of the plastic electrical box.

In most cases, the center pin of the power supply jack will be positive (+) and the outer barrel will be negative (-). You may wish to use a volt meter to confirm this before proceeding to internally wire the power supply for your MMM-DIY.

Use a soldering iron to connect a wire to the negative terminal of the installed female power jack. Black is a customary color for negative. Strip the other end of the wire. Connect the other end of the wire to the Arduino/ RAMPS board by inserting it into the screw-down lug socket marked negative (-) next to the 5A power input. The 11A power input will go unused.

Use a soldering iron to connect a red wire to the positive terminal of the installed power jack. Connect the other end of this wire to one of the two terminals on the toggle switch. The switch should be a simple SPST on-off type switch.

Solder one end of a new red wire to the other terminal of the switch. The other end of this wire should be connected to the Arduino/RAMPS board by inserting it into the screw-down lug socket marked positive (+) next to the 5A power input. Again, nothing should be screwed into the two open sockets for 11A.

10.2 Stepper motor jumpers & motor drivers

The stepper motor drivers are miniature circuit boards that allow the low power signals from the microcontroller to energize the coils of the stepper motors with controlled high-current. The Allegro A4988 stepper controller is shown in figure 10.2A. Stepper drivers can drive motors in full steps (the steps you feel the motor's shaft spring between if you turn it by hand.) They can also drive a motor in factional steps through 'microstepping'.

Enable microstepping for each motor driver by placing three PCB jumpers across the pairs of pins located below each stepper driver, see figure 10.2B. They are quite small and will require a small clamp and a steady hand. The locations of the jumpers on the board are shown in green in figure 10. Each of three jumpers are put in place before the motor driver boards are installed.

figure 10.2A

motor driver

figure 10.2B

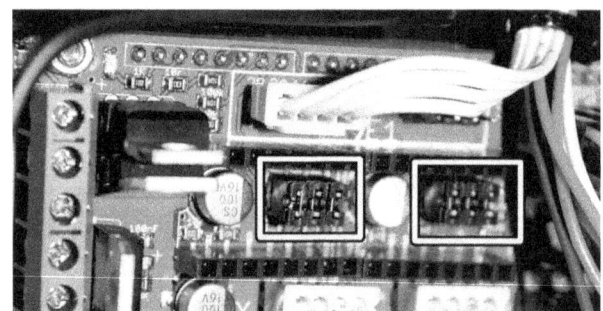

Once the jumpers have been inserted into the board, install each of the four motor driver boards. The board locations are showing in shaded green in figure 10. Note: The boards must be inserted in the correct direction or they will burn-up when the system is energized! Looking at the board with the USB jack pointing to the LEFT, the stepper drivers should be inserted as showing in figure 10.2A, with the small (+) shaped silver adjustment pot on the RIGHT side of the board.

Never insert or remove components into the controller board when power or the USB jack are connected.

10.3 Stepper Motors

Each of the four linear stepper motors should be mechanically installed into the motion systems of the MMM-DIY. Each motor should have come with a 4-pin connector that has bare wires at one end. The wire colors are red, green, blue and yellow. This 4-wire stepper motor type is referred to as a bipolar stepper motor. The electrical schematic is shown in figure 10.3. Notice that, electrically, the motor is two separate coils. Regardless of the color of the wires you are provided, you can always use an ohm meter to determine which pairs of wires are associated with a coil. With both coils connected to a driver, a wrong guess will mean the motor will only buzz rather than running. Reverse one pair, and the motor will run. The desired direction of running can be changed by reversing both pairs of wires.

The wires from each motor are not long enough to reach all the way back into the control box, so they will need to be extended. You can splice each wire to a new length of wire of the same color by first stripping the ends, then twisting them together. Finally, the wires should be individually soldered and the splice should be insulated with 2 cm of shrink-wrap tubing. The tubing is cut to length, slipped over one end of the wire and positioned. It is then secured by heating with a lighter flame. That is the nicest sort of splice, but you can also just use electrical tape if you are going for that hacker aesthetic. ;->

In order to connect the four wires from each wire to the controller board, they must be put into a pin header connector.

Crimp a female pin header crimp connector onto the stripped end of each wire. You may solder each if you wish, but it is not necessary.

Holding each pin by its crimped end with a pair of needle nose pliers, push each female pin into the open socket slots of the 1x4 pin header connector housing.

The wires should be inserted into the pin header connector housing in the following order: Red-Yellow-Blue-Green. Refer to figure 10 in order to see where and in which direction each of the four motor connectors should be attached to the controller. Note: Connecting the motor's connector backwards will cause the motor to run in the opposite direction. This may be useful to know at some point!

10.4 End of range switches

The end-of-range switches provide your MMM-DIY with a physical way of confirming its start position. The device can run its motors in the direction of 'home' until the motion carriage depresses the switch for that axis. The software then knows it can reset its position variable to zero for that axis.

Each end of range switch should have a pair of wires attached to the C (common) and the NC (normally closed) pins, as shown in figure 10. Each pair of wires should be terminated with a 2-pin wire jumper pin header connector. Use the same method described for connecting pin header connectors in step 10.3, but this time just use the 1x2 pin header connector housings. The end of range switches are then connected to the controller board as shown in figure 10.

10.5 Fan

Connect the fan to the board as shown in figure 10.

figure 10.3

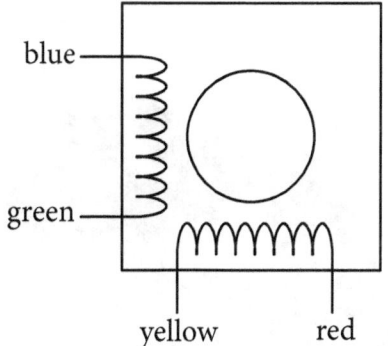

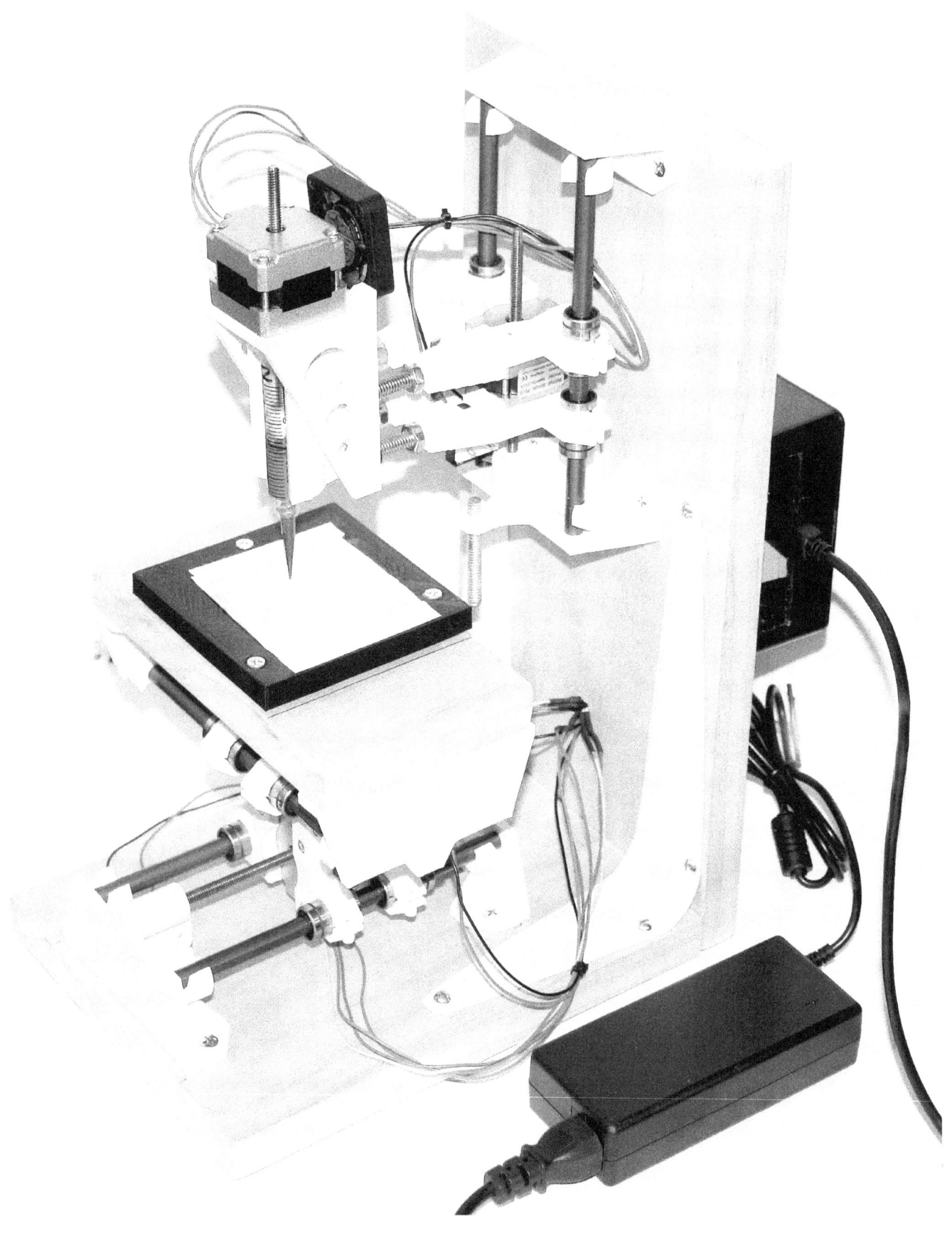

figure 10B - Completed Mini Metal Maker DIY Hardware

11. Firmware

The Mini Metal Maker DIY, like any 3D printer, is essentially a robot capable of executing commands. The commands, in this case, will come from an external desktop or laptop computer, and will be carried out by the Arduino microcontroller installed in the MMM-DIY.

In order to make sense of the commands, the Arduino microcontroller must run certain software that defines its behavior. The software that runs directly on the Arduino microcontroller is referred to as firmware, and determines everything about the controller's behavior. This includes the protocols it uses to connect to a PC, and the ways it drives the stepper motors when printing.

The firmware for the Mini Metal Maker is based on a popular open-source firmware called Sprinter, created by Kliment and based on firmware by Tonokip. The Sprinter firmware is licensed GNU GPL v3. You can read more about the Sprinter firmware on the REPRAP wiki at http://reprap.org/wiki/Sprinter. To read more about the GNU GPL v3 license, you can visit this page: http://www.gnu.org/licenses/quick-guide-gplv3.html

The following steps will guide you through connecting to the Arduino microcontroller and downloading the firmware software onto it.

Once this is done, the printer can be used by various different 3D printing software packages. This manual will walk you through using a set of the most popular free open source software programs.

11.1 Installing Arduino software

Arduino is an open source hardware and software platform. Essentially this means that anyone can create and sell the Arduino hardware without paying royalties and anyone can develop and distribute the Arduino software for free without infringing copyright.

If you purchased an Arduino board and RAMPS shield pre-assembled, then it most likely comes with a bootloader already installed on the board. If you don't know what that is, don't worry. If you don't have a bootloader, you will know soon enough. It's just another step to install one.

Install the Arduino development environment on your PC or Mac by going to the Arduino download page:

http://arduino.cc/en/main/software

Just download the most current version and follow the prompts to install it on your computer. Arduino hardware and software are very cool because they are open source and because they were developed to be user friendly tools to introduce beginners to the world of microcontrollers and programming. They have a real foothold in the world of DIY electronics, and are well supported around the world by programmers and hardware developers.

11.2 Modified Sprinter firmware

Once you have installed the Arduino software on your computer, start the program. It will load with a blank white "sketch" window. Programs in Arduino are called sketches.

Go to File -> Open

Browse to your folder for the Mini Metal Maker DIY instructions. You will open a file called Sprinter.pde located in the MMM DIY directory, as seen below.

figure 11.2A

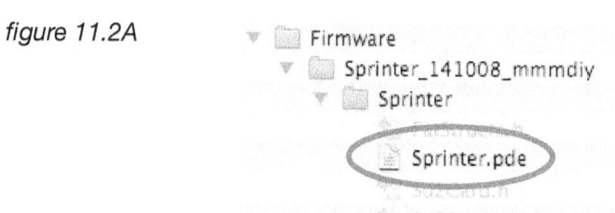

Once you select Sprinter.pde, the entire project will load in the Arduino development environment, see figure 11.2B. You will notice that there are many different tabs of code across the top of the sketch window. You do not need to actually know what any of these do, but it may be of some interest to you. It is all explicitly coded out and commented, so feel free to dig into it. You can also easily make changes to the code from any of these windows, so be careful not to mess it up and then press "Save."

Hint: Most of the specially configured lines of code reside in the library named Configuration.h

If you mess it up, just re-download it from the following: github.com/MiniMetalMaker/MMM-DIY

11.3 Loading firmware onto the MMM-DIY

With the firmware loaded into the Arduino software, connect your Mini Metal Maker to a USB port on your computer with a standard printer cable.

Note: The USB port will power the arduino card, so you do not even need to have the MMM-DIY connected to your 12V power supply for this step.

Try uploading the Sprinter firmware to the card by pressing the upload button at the top of the Arduino window. (See the button circled in red in figure 11.3A)

If the Arduino software fails to connect to your Arduino, check under Tools / Serial Port to see that the software is pointed to the proper USB port. You may just need to switch the USB port that is selected. See figure 11.3A.

If all goes well, the upload will begin, a set of LEDs will come to life and brightly blink on your Arduino board, and a few seconds later, you will be presented with the message "Upload Complete" on the Arduino software.

If you are having issues because your Arduino does not have a bootloader installed, the following the tutorial describes how to install one:

http://arduino.cc/en/Hacking/
Bootloader?from=Tutorial.Bootloader

11.4 Alternative firmware

If you are interested in exploring different types of firmware for your 3D printer, check out the firmware section of the REPRAP wiki at the following URL: http://reprap.org/wiki/Firmware

figure 11.2B

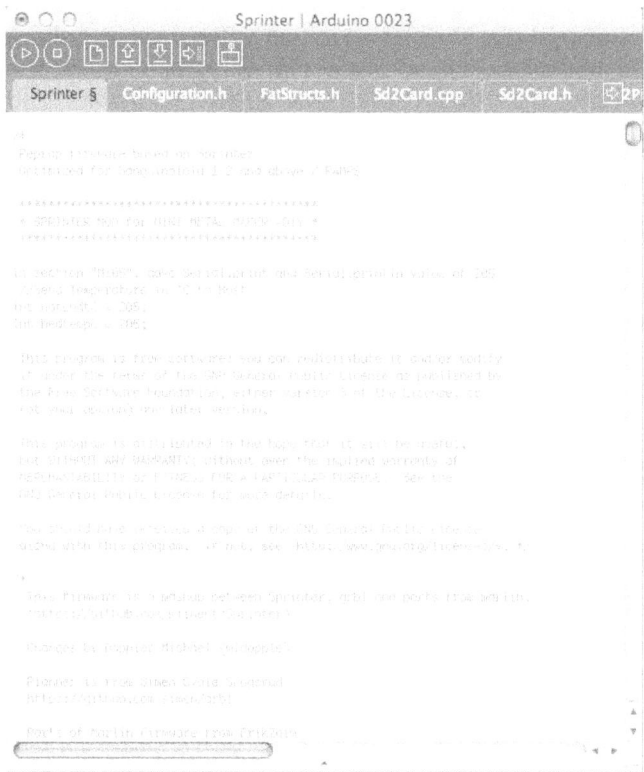

figure 11.3A

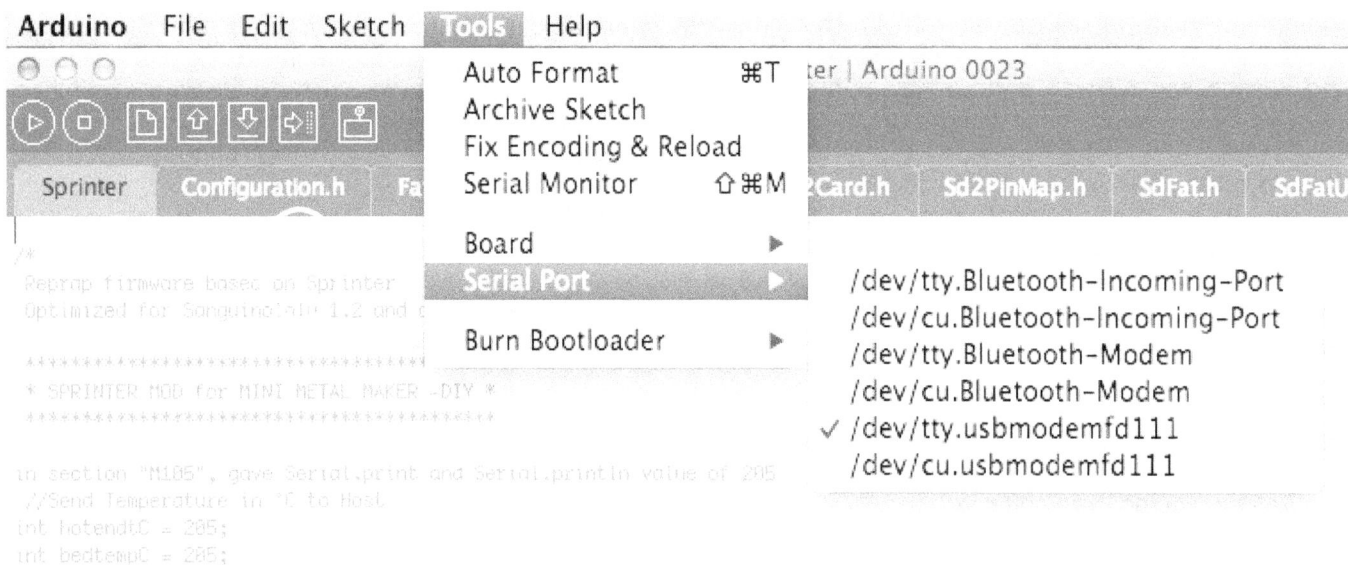

12. Software Toolchain

A 'toolchain' is a description of the software programs are used through the entire process of 3D printing with a given printer.

12.1 Overview

The process of creating a physical object with the Mini Metal Maker DIY involved four main categories of software: 3D modeler, slicer, host, and firmware. Figure 12.1 shows their relationship in the process. In each case, the program creates a specific file type that is then opened by the next program in the chain, processed and saved out again as a new file. The file type that is created is shown above the arrow connecting each item in the chain.

12.2 3D Modeling

There are many different software packages for creating 3D objects that can be 3D printed. Any modeling software that can be used for standard 3D printers will work for the Mini Metal Maker DIY. There are, however, some additional considerations for items you wish to print with clay. These will be discussed in section 15.

The following are some free 3D modeling programs:

Autodesk 123D
Tinkercad
3D Tin
Blender
FreeCAD
OpenSCAD
Sculptris
Sketchup
3D Model To Print

Most of these will run on either Mac or PC, and a few, most notably Tinkercad, are designed to run directly within a web browser. These programs are great introductory tools, and each have specific strengths and weaknesses. Any of them are a great place to start.

For more detailed or technical projects, you may require higher end modeling software.

The following are professional modeling programs:

3DS Max
Autocad
Cinema 4D

Lightwave 3D
Maya
Photoshop CC
Rhinoceros
ZBrush
SolidWorks
Netfabb

12.3 Slice

A 3D printer constructs objects by moving an extruder head in 3 axis while drawing with a liquid material. The printer physically coordinates this process by moving its motors at different speeds at different times throughout the print. The moves that the motors make are directed by a set of digital instructions. These instructions are typically referred to as the 'toolpath' and are most often saved in the file format .gcode.

The .gcode takes into account things about the printer such as it's bed size, the kind of material that is being printed, and the speed at which the motors can operate. It is very hardware specific.

A 3D object file, in contrast, is an abstract definition of an object. It describes the size and shape of an object, but carries no information about how it might be constructed.

The process of transforming a 3D object file into a toolpath for a specific printer is commonly called 'slicing.' This is because most 3D printer tool paths are organized as successive layers that are to be printed. Cutting something into lots of layers can be thought of as slicing. That said, slicing is just a word for toolpath creation.

There are several stand-alone options for slicing objects. The slicing process may also be carried out by a 'host' program designed to also connect with a printer and to direct the printing process. Some examples of different programs that generate toolpaths are the following:

Cura
Repeiter-Host
Reprap host software
RedSnapper
SkeinForge
Slic3r
SuperSkein
YARRH
X2SW

figure 12.1

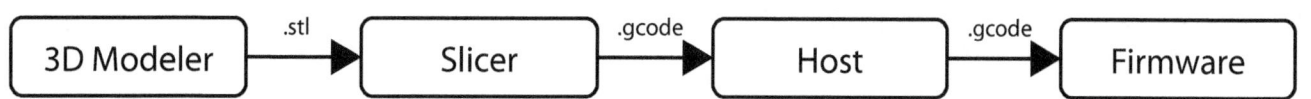

Figure 12.6

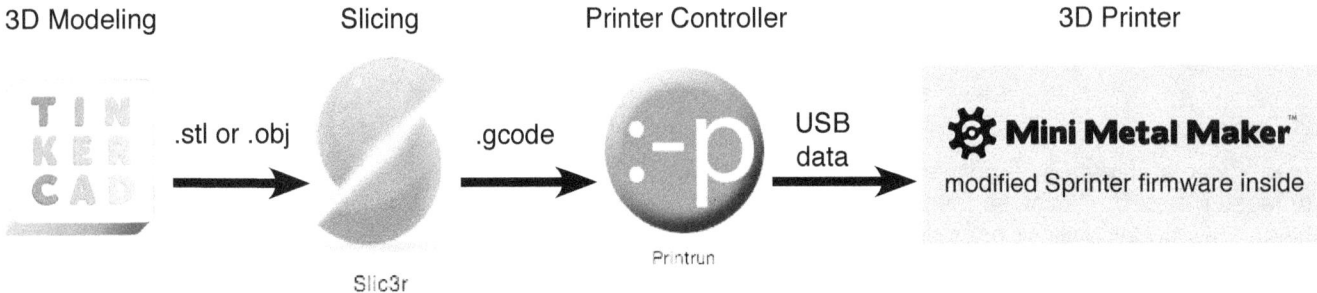

3D Modeling   Slicing   Printer Controller   3D Printer

.stl or .obj   .gcode   USB data

Slic3r   Printrun   modified Sprinter firmware inside

12.4 Host

Typically, there is a program that allows a person to connect to a 3D printer and to set-up and start the print process. This program often includes a graphical user interface and is usually referred to as a 'host. ' It might also be called the 'controller.'

The main purpose for the host software is to pass the g-code toolpath to the 3D printer. The host also typically provides controls for calibration and the changing print material. It may also include manual controls for driving the printer's motors for diagnostic purposes.

Some common open-source host programs include the following:

CNCGcodeController
EMCRepRap
GCodePrintr
GCode-utils
OctoPrint
Printrun (AKA Pronterface)
A larger list of slicing and host software can be found on the RepRap wiki at the following:

http://reprap.org/wiki/Toolchain

12.5 Firmware

As described in section 11, the firmware is the software that runs inside the actual 3D printer. Firmware is usually very low-level software, meaning it is directly in touch with the hardware of the system. In a 3D printer, the firmware generates signals from the microcontroller's output pins, which connect to the stepper motor drivers.

Most firmware for 3D printers operates as a sort of translator, which receives higher level commands in the form of 'g-code' and translates those commands into action. The firmware converts abstract commands like, "move 2.5 cm in the X-Direction" to actions like sending a chain of 350 output pulses to pin number 26 of the Arduino microcontroller. Presumably, in this example,

pin 26 would be connected to the step-input pin of the X-motor's stepper driver card.

There are many different types of firmware in use for 3D printers today. Some are proprietary and some are open-source. If you are interested in exploring different flavors of firmware for your 3D printer, visit the RepRap wiki at www.reprap.org.

12.6 Sofware setup

Since the MMM-DIY is at its core a do-it-yourself project, we support experimentation and do not want users to be locked into any particular toolchain or control software.

Figure 12.6 presents one toolchain that we have extensively tested with the Mini Metal Maker.
As an experimenter, you may choose to work with different modelers, slicing software, and control software altogether. There are also different flavors of firmware that work well with the Arduino/RAMPS hardware, Marlin being the most popular. We chose to use Sprinter firmware because it is easy to setup and to hack from the Arduino development environment.

The newest versions of all of the following applications are available for download online from various different developers. There is a list of quick links on the back cover of the manual to make it easy to find all of the downloads you will need.

12.6.1 Download Slic3r

Go to the following address and download and install the program Slic3r: *http://slic3r. org/download* When the application starts, it should look much like shown in figure 12.6.1.

12.6.2 Download MMM Config for Slicer

The next step is to download a special configuration file that we have created specifically for the Mini Metal Maker. This config file sets up Slic3r to produce .gcode for the Mini Metal Maker. Go to the following address:
github.com/MiniMetalMaker/MMM-DIY

This is the online repository for all of the software developed specifically for the Mini Metal Maker. Click on "Download Zip" and you will download a folder called "MiniMetalMaker-master" to your computer.

Unzip the file and move the entire folder to your documents directory or somewhere you can find later.

Now, in Slic3r, go to File and choose *Load Config...* You can find the Slic3r config file inside the MiniMetalMaker-master folder as shown in figure 12.6.2A.

The configuration file name starts with a date in yymmdd format. Choose the newest config file if there are more than one. Once the config file is loaded into Slic3r, it will appear in the settings as shown in figure 12.6.2B.

Note: If you can't see which settings are loaded in slic3r, go to the top of the screen, click Slic3r, choose 'Preferences...' and make sure the 'Mode:' is set to 'expert.' Then the settings will show.

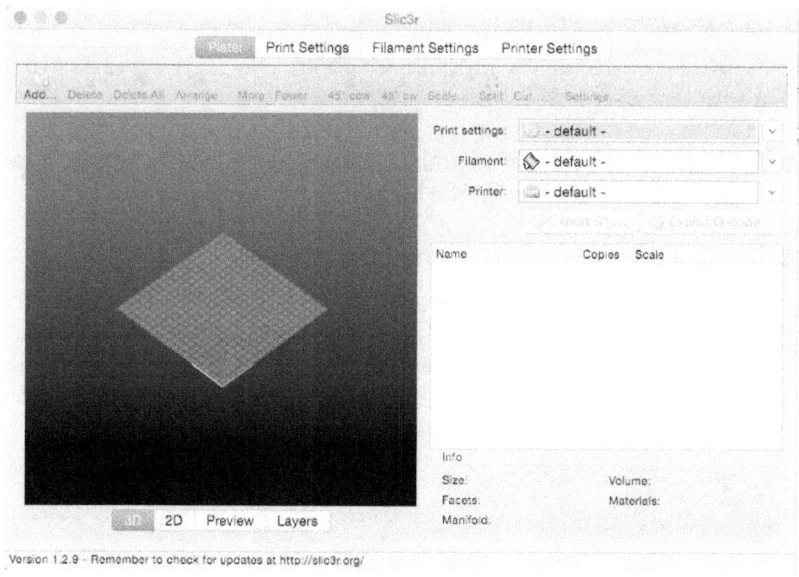

Figure 12.6.1 The Slic3r application when first opened after downloading.

Figure 12.6.2A Locating the MMM configuration file for Slic3r.

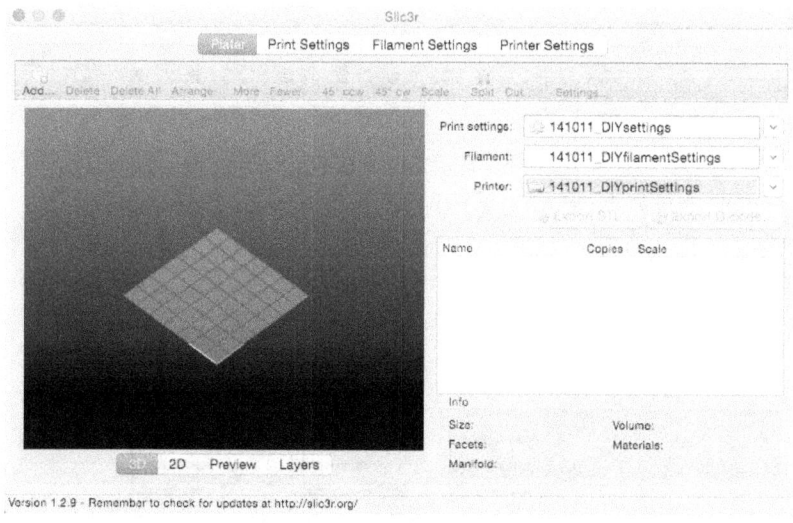

Figure 12.6.2B The MMM config file successfully loaded into Slic3r.

12.6.3 Download Printrun

Printrun is a powerful free program for controlling 3D printers. Go to the following site to download and install Printrun aka Pronterface: *https://github.com/kliment/Printrun*

Once Printrun is installed on your system, you will need to make a few adjustments to its configurations in order for it to work well with the Mini Metal Maker.

12.6.4 Configure Printrun for the MMM

Start Printrun (also known as Pronterface). Figure 12.6.4A shows how Printrun should look:
From the menu at the top, click Settings and choose Options. With the 'Printer Settings' tab selected, make your settings match those shown in figure 12.6.4B. Once this is done, all of your software is configured.

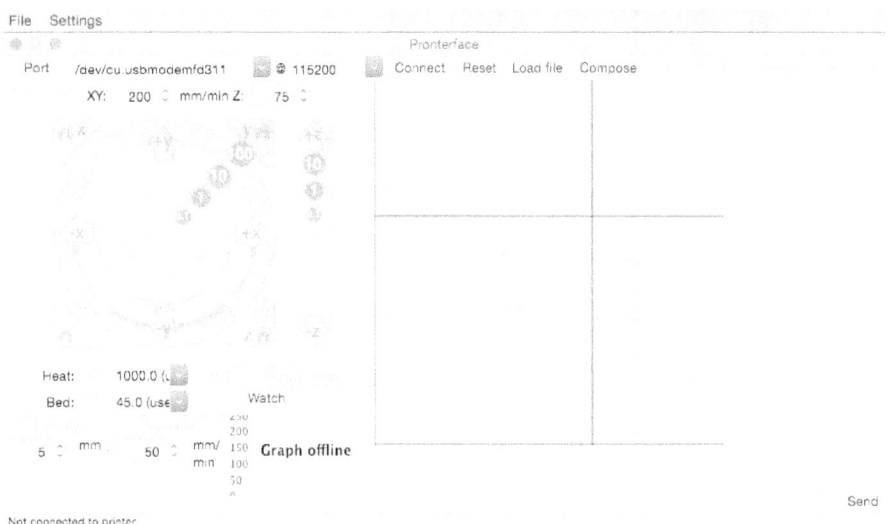

Figure 12.6.4A The Printrun (Pronterface) interface.

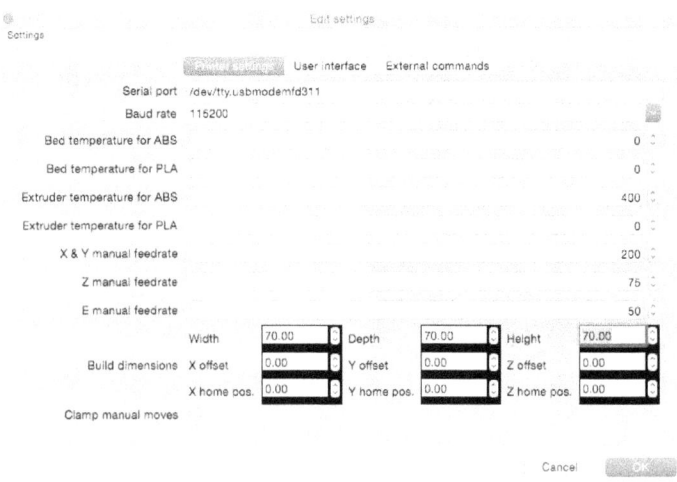

Figure 12.6.4B Printer settings in Printrun.

13. Metal clay

Metal clay is a mixture of particulate metal and a combustible binder, typically mixed with water or some light solvent. Metal clay can be sculpted by hand as one would use clay. The material is then allowed to dry. Once dry, it is heated to a high temperature at or near the melting temperature of the particulate metal in the clay. The binder material effectively burns out while the metal particles bond to each other. The over all shape of the object is maintained during this process, though some dimensional shrinkage occurs. The finished object is composed of solid metal, which can be polished or machined to any finish.

Metal clays exist for a variety of metals, including precious metals such as gold, silver, and platinum. More common alloy metals such as bronze, brass and nickel-tin. There are also metal clays composed of copper, iron, and aluminum.

Metal clays can have extremely different physical properties depending on the metal, the firing process used, and the heat treatment of the item after firing.

Metal clays are most often used by jewelers or sculptors who work with ceramics to make beads or fused glass. Firing metal clay requires a small electric kiln. The finishing work for metal clay is very similar to the process used to finish metal jewelry, including filing, sanding, polishing as well as the use of various metal finishing compounds.

There is an entire world of metal clay artistry and technique. If you are interested about learning more about working with this material, there is a nice introduction at the following website:

http://www.metalclayguru.com/

13.1 Suppliers & types

Clay consistency is of extreme importance when extruding with the Mini Metal Maker. At one extreme, if a clay is too thick, the extruder will not be strong enough to push it through the extrusion tip. At the other extreme, if the clay is too wet, extrusion will be easy but your object's form will collapse or slump out of its desired shape.

There is a formulation of bronze metal clay available online at www.minimetalmaker.com that will work with the commercial Mini Meatl Maker. The syringe for the commercial Mini Metal Maker is too large to be extruded by the DIY version. You could purchase the clay online and refill the smaller syringe sizes, or else experiment with mixing your own.

Metal clay of the correct consistency can be obtained by mixing water with metal clay in powder form. It can also be obtained by further wetting pre-mixed wet metal clay, but this approach is more difficult and is often not as homogenous as mixing from powder.

I recommend beginning with the lower cost bronze or copper type metal clays. The following is a short list of metal clay suppliers and manufacturers:

Rio Grande
http://www.riogrande.com/

Metal Clay Supply
http://www.metalclaysupply.com

Cool Tools
http://www.cooltools.us/

Metal Clays.com
http://www.metalclays.com/

13.2 Mixing method

Tools you will need:
- small gem scale with fractional gram resolution
- plastic film canisters
- plastic syringe for dispensing water
- round chopsticks for mixing
- small spreading knife
- teaspoon

Supplies you will need:
- powdered metal clay
- glass of water
- paper towels
- empty 5-10 mL syringes (see section 13.3)

It is important to do your clay mixing in a clean environment on a clean surface. The smallest chips or granules of foreign material can ruin a batch of clay because it will clog the tip of your extruder. Also, pet hair can be a serious problem. Seriously.

Our dog, Kita

Begin by removing the lid from a clean and dry empty film canister. Place the canister on the gem scale and zero the scale.

Next, add 15g of powdered metal clay to the canister using the spoon. Be very careful not to spill clay outside of the canister onto the scale. If you overshoot the mark,you can use the spreading knife or the end of the chopstick to lift small portions of powder free from the canister.

Once you are satisfied that you have an accurate 15g of powder, use a syringe to add 3.25 g of water to the canister. This will be very close to 3.25 cc as read on the syringe, but measurement by weight is more accurate than by volume markings.

Once your total weight reads 18.25g, use a single round chopstick to mix the clay in the canister. Use firm slow motions, being sure to sweep the bottom and the inner edges of the canister. Mix for a minute or so, then turn the chopstick against the inner wall of the canister while removing it.

Replace the cap and let the clay mixture sit overnight at room temperature.

Give the clay a final slow stirring upon opening the canister.

Finally, use the small spreading knife to scrape the clay from the canister and deposit into the open end of an empty syringe. The process described should yield around 4.25mL of finished metal clay of a consistency that will work with the MMM-DIY. The consistency will vary depending on the brand and type of clay you selected. This is one area that will require some experimentation on your part, especially if you are printing challenging structures that are 20+ layers tall or have overhangs.

figure 13.3A

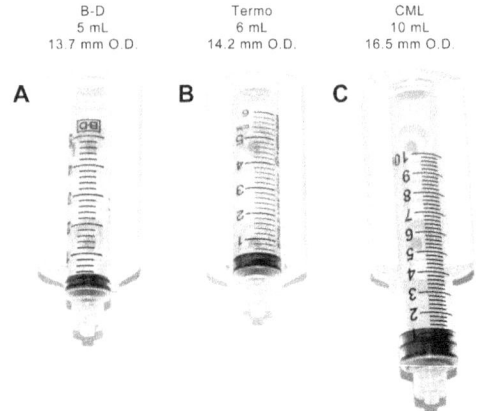

13.3 Dispensing syringes & tips

The 3D print models included with the DIY-MMM include three differently sized SyringeMounts, and one set of objects that can be used to create a custom-sized SyringeMount. Figure 13.3A shows the three syringe mount sizes along with a the particular size and brand of syringe that fits each.

For the purposes of this tutorial, we will be teaching the use of SyringeMountB with the 6mL syringe by Termo. In order for the MMM-DIY to push clay in a syringe, the plastic plunger of the syringe must be cut short as shown in figure 13.3B.

In order to extrude thin lines, it is important to have a consistent small diameter dispensing tip. We have had the most success with conical dispensing needles like the type supplied by CMLsupply.com. Figure 13.3B shows a typical plastic dispensing tip in the 22 gage size.

Any of these syringes can be easily purchased online through eBay or Amazon. Additionally, the website CMLsupply.com has a variety of syringes in stock.

figure 13.3B

Prepare the syringe by clipping the plunger near the piston, leaving a 3 mm height of the plunger's stalk.

figure 13.3C

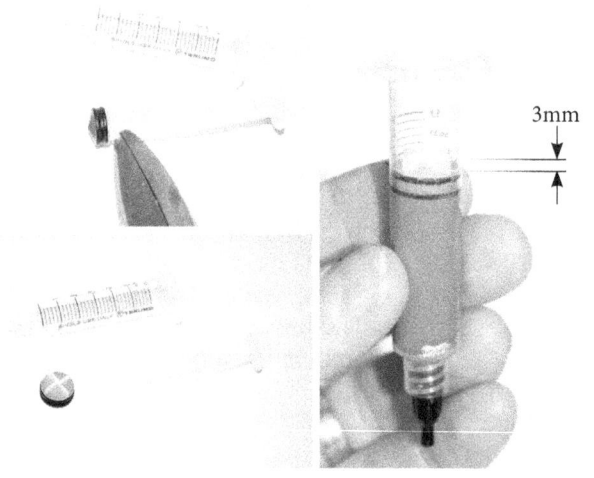

14. Machine Setup

14.1 Set motor driver current limiters

Looking at the miniature motor driver boards that are plugged into the RAMPS card, adjust the small silver pot for each with a miniature plastic screwdriver. Figure 14.1 shows the appropriate setting for the pot.

figure 14.1

Current limiter settings

motor driver

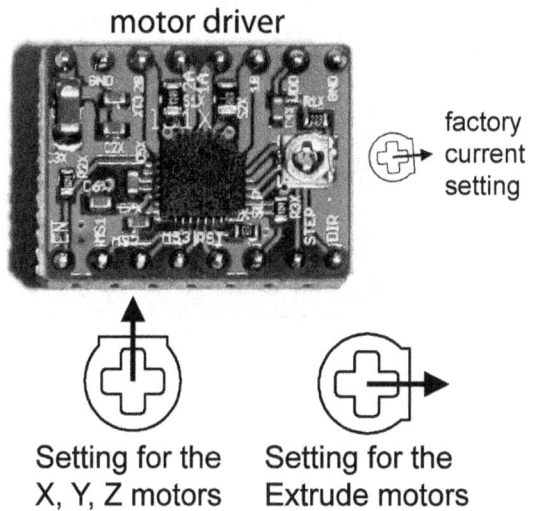

factory current setting

Setting for the
X, Y, Z motors

Setting for the
Extrude motors

14.2 Test X, Y, Z direction & ranges

Start Printrun, AKA Pronterface. Click connect, and wait for the 'connected' message to appear.

Press "HOME" button. This is the small button in the lower left of the 'bullseye' with an icon that looks like a small log cabin. See figure 12.6.4A. The motors should all run in the toward the endstops for each axis, and stop when they reach endstop switch. If a motor runs in the opposite direction than you expected, make a note of it. You can simply reverse the motor by reversing the direction of the motor's connector in the circuit board later.

If the motor crashes into the endstop without stopping, you may need to check your connections to the switch. The switch may be connected to the wrong pins on the board, or you may have wired it improperly. Check again with figure 10.

Next, press the Extrude button, and check to see that the extrusion motor runs and in the proper direction to press the piston into the syringe. Again, take note if it runs in the wrong direction. You can reverse the direction by rotating the motor's connector to the circuit board 180 degrees.

Next, we are going to check the other end of the range. By default, the MMM-DIY is set to have a 7 cm range in the X and Y directions and a 4 cm range in the Z direction. You are free to extend these ranges within the firmware (using the arduino development environment and re-uploading the firmware) but this default is safe.

The default setting should let you manually control each axis' movement away from the zero position until the maximum distance is reached, and then it will stop. You will still be able to move toward the zero position, but no further than the specified limit.

14.3 Loading the clay syringe

Turn the threaded shaft of the extruder motor CCW until the plunger is all the way up against the underside of the motor.

Place a syringe of clay against the open side of the syringe holder. Firmly press the top and then the bottom with the thumb, holding the extruder module with your fingers. The syringe will pop firmly into place, and will be centered below the plunger (figure 14.3).

Finally, turn the threaded shaft of the motor CW until the plunger is in contact with the syringe piston within the syringe's tube (figure 14.4).

14.4 Aluminum foil build surface

Use thin strips of paper tape around the edges of a square flat piece of aluminum foil. Foil is an excellent build surface for metal clay because wet clay will stick to its surface, and will release once dry. It is also inexpensive and stiff enough to be removed from the machine while the clay is still wet.

figure 14.3

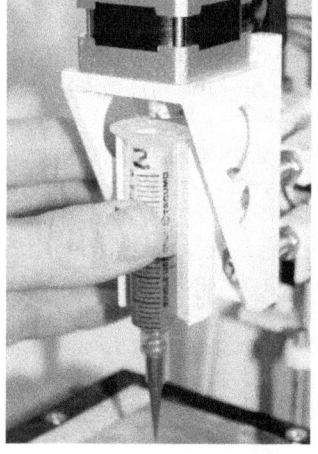
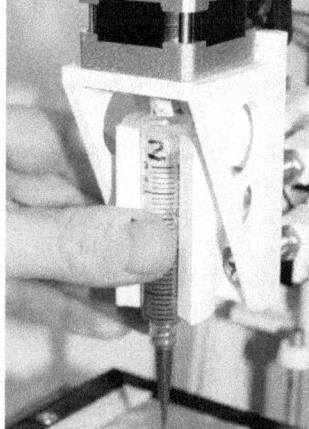

figure 14.4

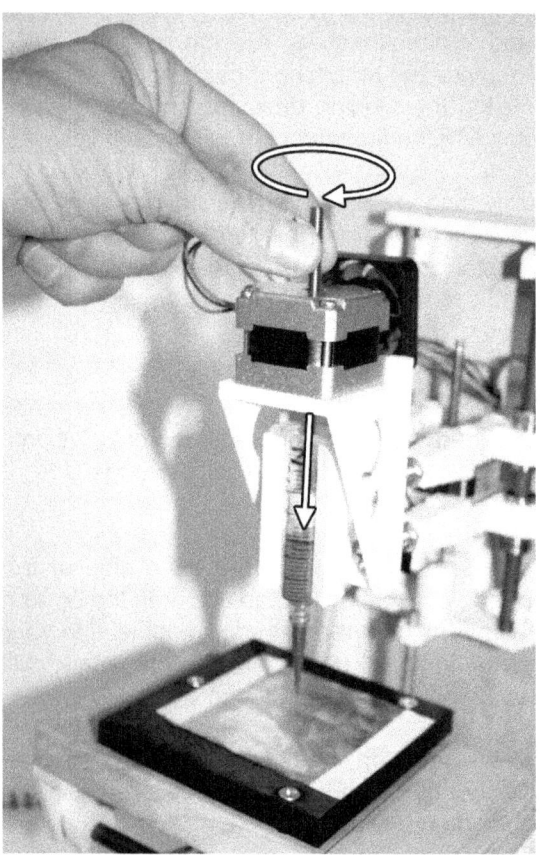

14.5 Setting the Z-axis zero position

With the printer connected, start Printrun AKA Pronterface and click "Connect."

Press the Home button to move all axis to zero position.

Manually drive the Z-axis up so that there is enough room between the end of the syringe and the build plate to attach a plastic dispensing tip, like the one shown in figure 13.3B.

With the tip in place on the syringe, there should be some clearance between the tip and the build platform.

Turn the Z-adjustment screw with your fingers (see figure 14.5) until the screw is almost touching the endstop switch for the Z axis. Press "Z-Home." There should now be a visible gap between the dispensing tip and the build platform. Turn the Z-adjustment screw again to lower it slightly, then press "Z-Home" again. Repeat this process with small incremental turns until the dispensing needle tip is just barely touching the foil surface. The tip should be just barely in contact but still free to move when wiggled.

You will need to adjust the Z-height each time you change the plastic dispensing tip or the syringe. You typically change the tip for each new print you make because tips are ruined once the clay flow stops and is able to dry in their end.

14.6 Leveling the build plate

You will only need to level the build plate very rarely, so don't worry. This is the most detail-oriented step in the whole process!

You will need a phillips head screwdriver and some calm patience. This won't take long.

First, Press the Z-home button and adjust the Z adjust screw (see fig 14.5) if necessary so there is a small gap between the nozzle and the build platform.

Next, use the manual X & Y controls to drive the platform so the nozzle is near each of the three leveling screw positions. Use the screwdriver to turn the screw nearest the nozzle until there is a gap only big enough for a piece of paper to slide under. Do this for each of the three adjustment screws.

Finally, drive the nozzle to the center of the platofrm and confirm your leveling by seeing if you can pass a piece of paper under it.

Once the platform is level, you are ready to start experimenting with your first print!

figure 14.5 Z-adjustment screw

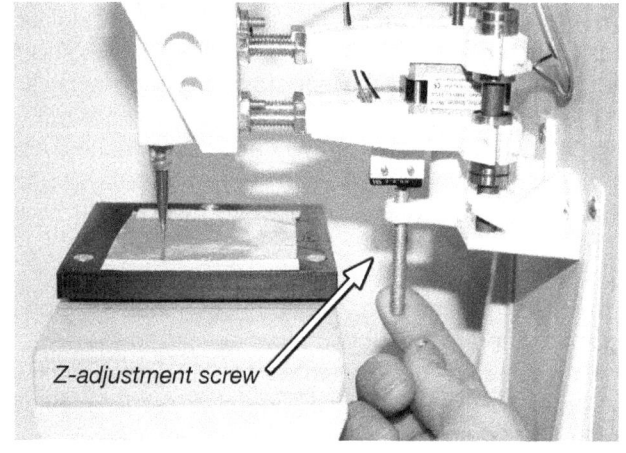

15. Printing a test object

With the Mini Metal Maker hardware and software set up and the build plate leveled, it is time to make a print. You may wish to start with one of the test objects included in the MMM-DIY directory that you downloaded. Test objects can be found in the folder named "Test Object." All of the test objects are in the .stl file format.

15.1 Prepare GCODE in Slic3r

The first step is to slice the .stl file into a .gcode file.

Begin by opening Slic3r. Next, load the .stl file into Slic3r by clicking the 'ADD' button in the upper left and choosing a .stl object.

Make sure that each of the preset configurations is set to the latest DIY_Slic3r_Config.ini. See section 12.6.2 Download MMM Config for Slicer for details on doing this.

At this point, simply press "Export G-code..." from the right hand side of the Slic3r window, and save the .gcode file anywhere you like. Look back to figure 12.6.2B for an image of the Slic3r interface.

15.2 Opening an object in Printrun

The next step is to load the object into Printrun so that it can be sent to the printer. Open your configured version of Printrun. See section 11.4 Software Setup for details about how to do this if you haven't already.

15.3 Priming the printer
Now it's time to put clay in your printer and get it ready for printing.

Turn on the printer and connect to it from Printrun. Printrun will say "Disconnect" instead of "Connect" when the printer is connected. The printer itself will not appear to be doing anything.

Next, load the clay. Refer to section 14.3 for details about how to do this.

Turn the extruder piston by hand to screw it down into the body of the cartridge. You will feel it get hard to turn once the piston is seated against the piston of the cartridge. Figure 15.3A shows this process.

Be sure to remove the end cap from the syringe if there is one. It is a good idea to keep the syringe capped for storage.

The first amount of clay at the start of a new cartridge is partly dried. Set the manual extrude distance to 5 in Printrun and press extrude. Wipe the clay free with your finger once it starts to flow from the end.

figure 15.3A

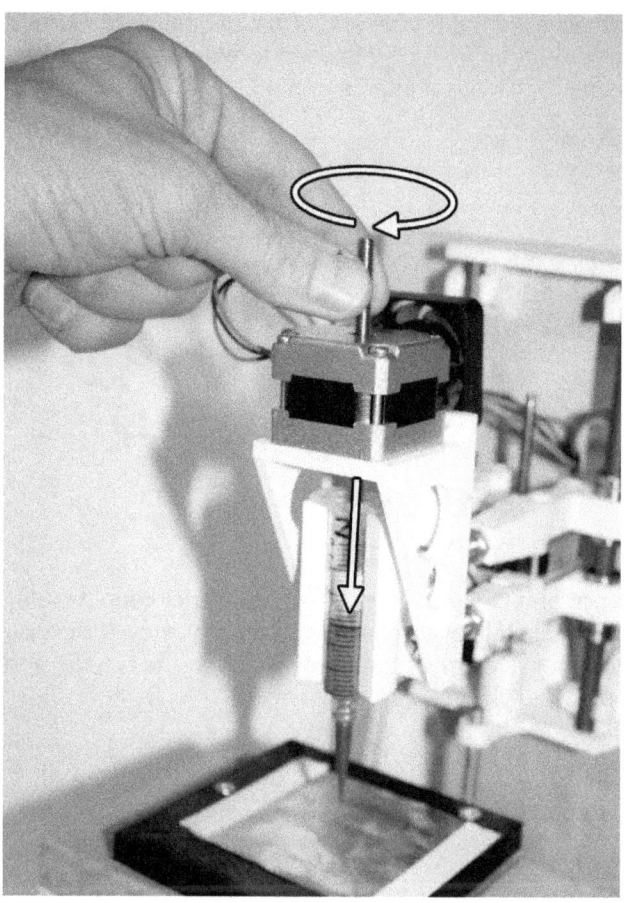

figure 15.3B

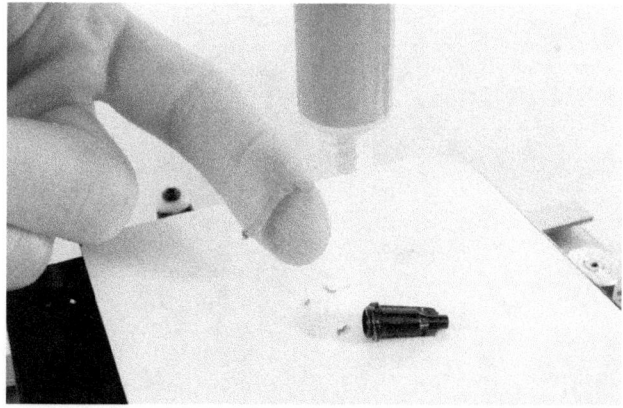

Once the clay is free to flow, place a fresh extrusion nozzle onto the cartridge. It twists into place finger-tight. You may need to raise the extrude head in order for the new nozzle to fit. Press the +Z 10 button in the manual controls of Printrun.

Press the extrude button a few times until clay has entirely filled the nozzle and is just beginning to extrude from the tip. Once clay is seen to emerge as shown in figure 15.3C, the printer is primed.

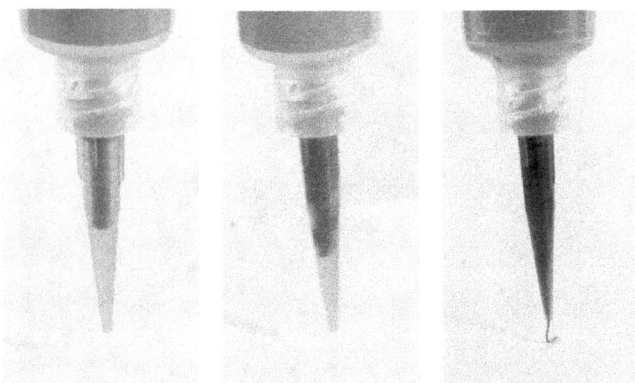

Fig. 15.3C

Important: Once primed, be sure to either keep the clay oozing slightly or else start a print right away. If the clay stops for even one minute, it will dry in the nozzle, and the nozzle will need to be replaced.

15.4 Adjustments

The most difficult thing to get right with the Mini Metal Maker is the clay extrusion rate. The best way to adjust the extrusion rate is actually by fine-tuning the "Extrusion multiplier" setting inside Slic3r. Figure 15.4A shows the location of this setting.

Increasing this number will cause the MMM to extrude faster. Reducing it will slow the flow of clay during the print. Very small adjustments to this number can have a noticalbe effect. For instance, a setting of 0.3 may be too fast, whereas 0.26 may be perfect.

I also highly recommend downloading and reading the complete user manual for Slic3r, available at this address: http://manual.slic3r.org/

Use the picture guides on the following pages to help determine the settings for your extrusion speed. You may wish to keep a log of your settings for different batches of clay.

figure 15.4A

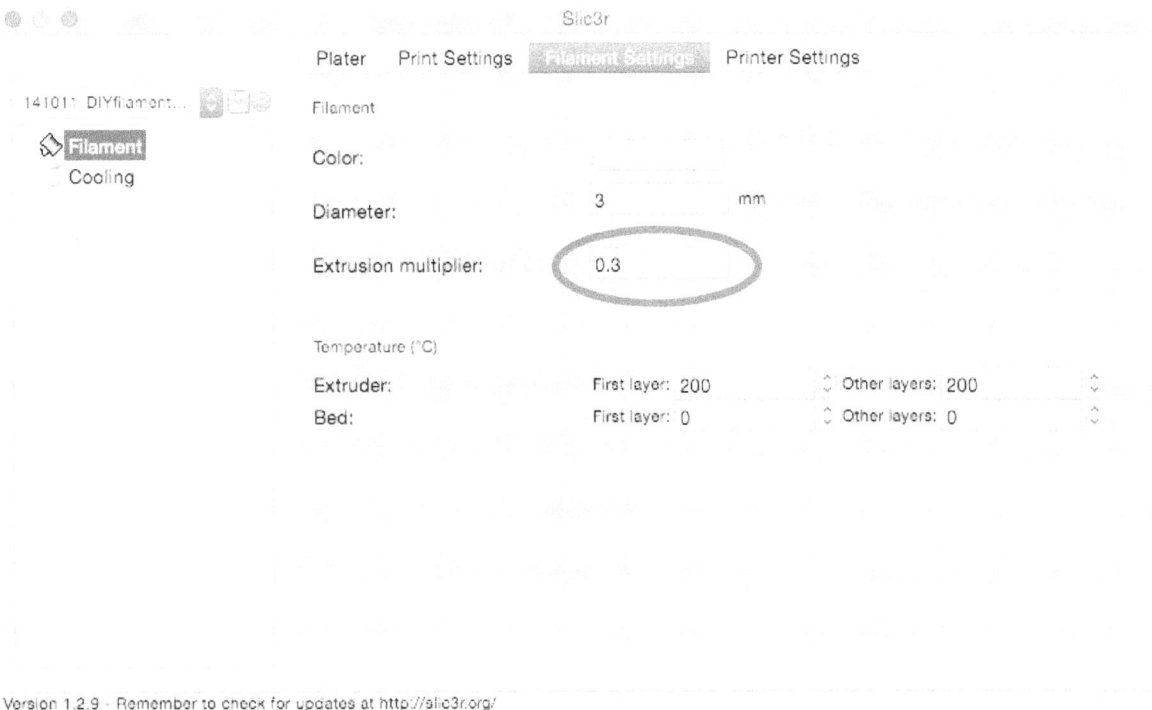

A pictoral guide to good clay extrusion

Under extrusion: Poor layer adhesion, gaps in the perimeters, sides look sunken-in. Lines of clay may appear very thin or may be seen to break during extrusion.

Setting 0.2

Good extrusion: Layers are even and are not bulged or sunken-in. Perimeters are unbroken. Each layer goes on flat with minimal buildup on the extrusion tip.

Setting 0.25

Over extrusion: Object appears bulged, corners undefined, layers seem to swell upward during printing, material builds up on extrusion tip. Printing seems muddy or messy.

Setting 0.3

Setting 0.5

16. Cleanup, Fire & Finish

16.1 Cleanup

Once a print is complete, allow it to dry overnight before handling. The print will warp and bend in unusual terrible looking ways while it is in the process of drying, but will be fine when it is totally finished drying. Don't watch your parts dry.

16.2 Kiln fire

Read the firing instructions that came with your particular choice of metal clay. Some metal clays require that you surround your part in activated carbon within a small stainless container inside the kiln. Figure 16.2A shows dried metal clay objects placed in a stainless container of activated carbon. Figure 16.2B shows a popular electric kiln used for metal clay, made by Paragon.

16.3 Finishing

Some of the finishing stages are most easily done directly to the clay before you fire. If a print has unsightly ridges, it is much easier to smooth the surface with a wetted finger as dried clay than trying to file it as metal later. A utility blade can be used to trim the edges of dried metal clay object before it is fired. Figure 16.3 illustrates the difference in texture and size of the same 3D print before and after the firing process. Note that the shrinkage along one of an object's dimensions can be as great as 15%. The exact shrinkage has a great deal to do with the type of clay and how long & at what temperature it is fired in the kiln.

Taking notes as to how you mix your clay and the exact firing settings of your kiln will let you produce repeatable results.

Figure 16.2A

Figure 16.2B

Figure 16.3

17. Reference

17.1 Slic3r Settings

The following screenshots show all of the settings in the program Slic3r, as configured to print with metal clay on the Mini Metal Maker DIY printer.

These settings work for a MMM-DIY using a 6mm Termo syringe (O.D. = 14.2mm) and copper clay by Hardar's mixed in the weight proportion clay:water 15:3.25.

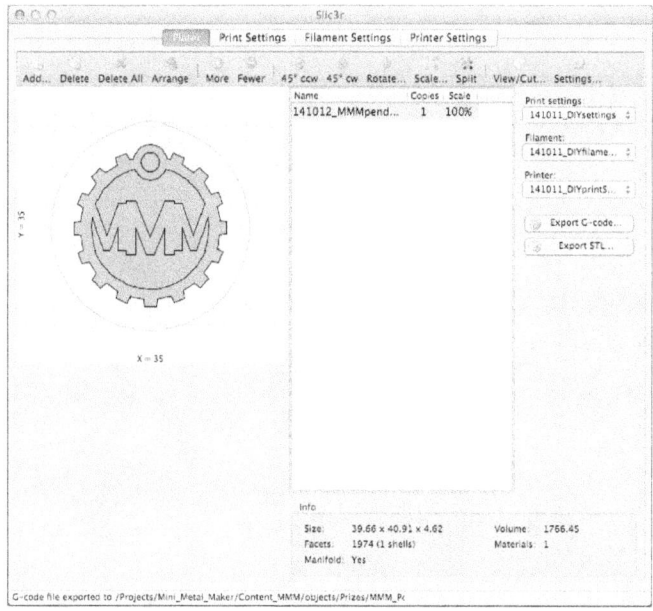

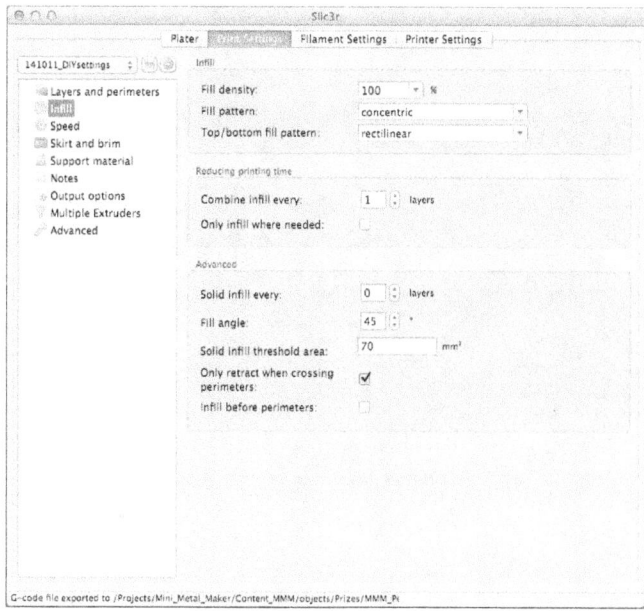

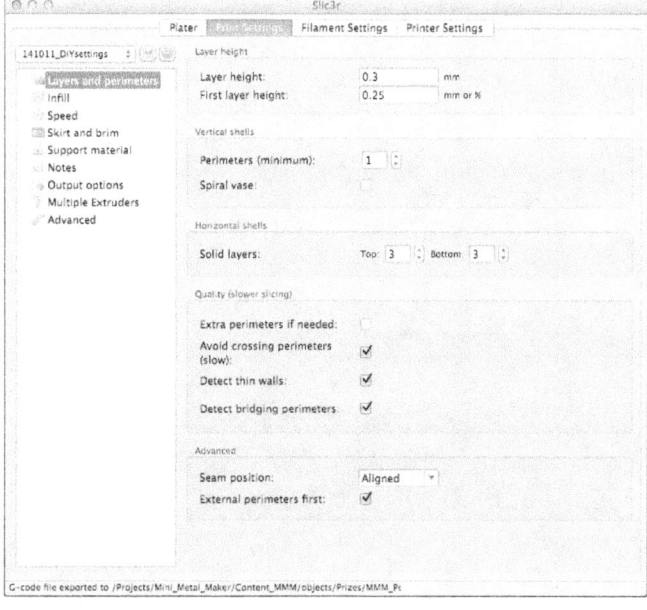

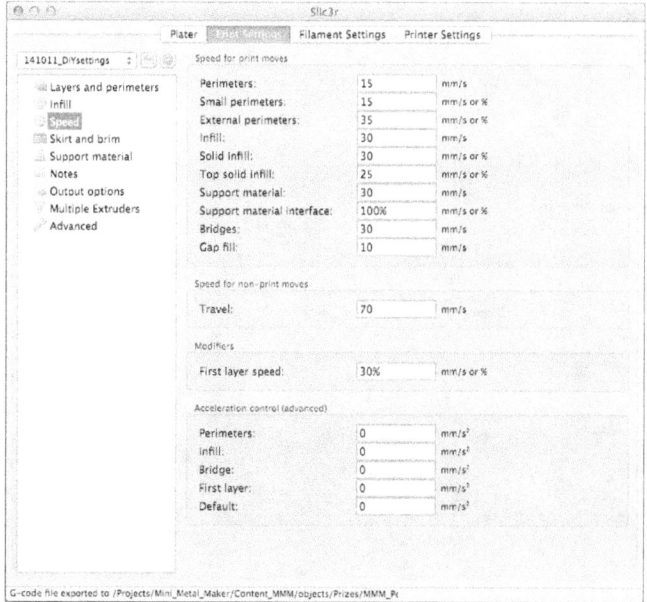

Panel 1 (top-left):

Slic3r

Plater | Print Settings | Filament Settings | Printer Settings

141011_DIYsettings

- Layers and perimeters
- Infill
- Speed
- Skirt and brim
- Support material
- Notes
- Output options
- Multiple Extruders
- Advanced

Skirt

Loops:	1	
Distance from object:	6	mm
Skirt height:	1	layers
Minimum extrusion length:	0	mm

Brim

| Brim width: | 0 | mm |

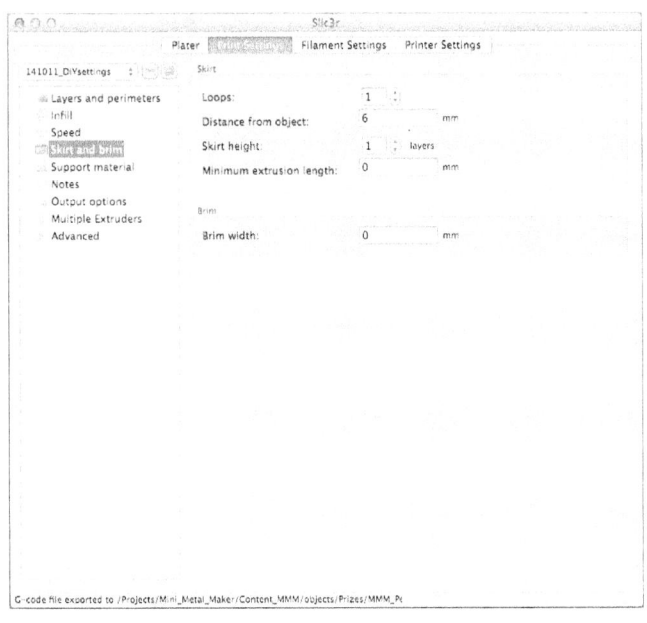

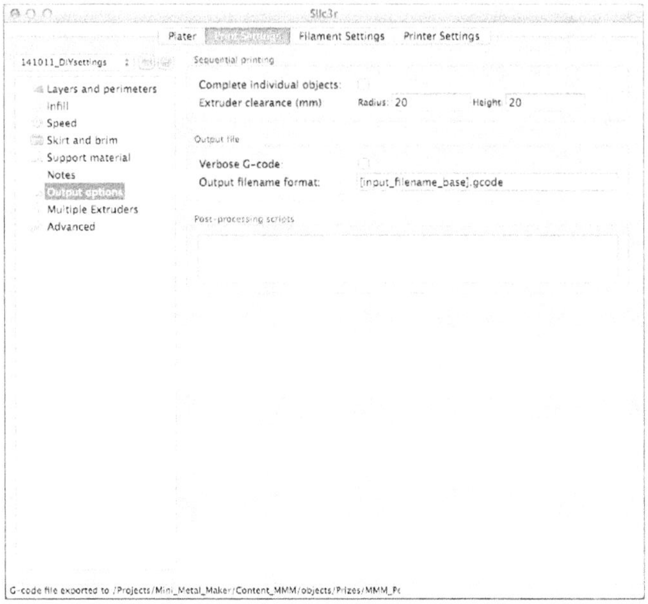

G-code file exported to /Projects/Mini_Metal_Maker/Content_MMM/objects/Prizes/MMM_Pc

Panel 2 (top-right):

Slic3r

Plater | Print Settings | Filament Settings | Printer Settings

141011_DIYsettings

- Layers and perimeters
- Infill
- Speed
- Skirt and brim
- Support material
- Notes
- Output options
- Multiple Extruders
- Advanced

Sequential printing

Complete individual objects:

Extruder clearance (mm)   Radius: 20   Height: 20

Output file

Verbose G-code

Output filename format: [input_filename_base].gcode

Post-processing scripts

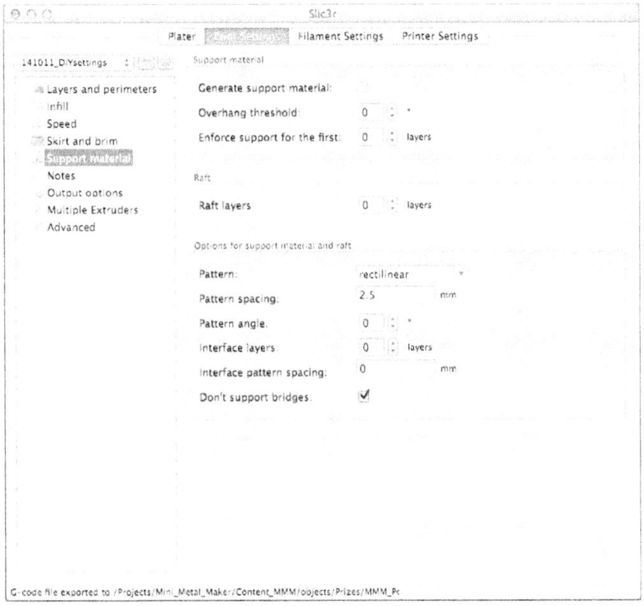

G-code file exported to /Projects/Mini_Metal_Maker/Content_MMM/objects/Prizes/MMM_Pc

Panel 3 (middle-left):

Slic3r

Plater | Print Settings | Filament Settings | Printer Settings

141011_DIYsettings

- Layers and perimeters
- Infill
- Speed
- Skirt and brim
- Support material
- Notes
- Output options
- Multiple Extruders
- Advanced

Support material

Generate support material:		
Overhang threshold:	0	°
Enforce support for the first:	0	layers

Raft

| Raft layers | 0 | layers |

Options for support material and raft

Pattern:	rectilinear	
Pattern spacing.	2.5	mm
Pattern angle.	0	°
Interface layers	0	layers
Interface pattern spacing.	0	mm
Don't support bridges.	✔	

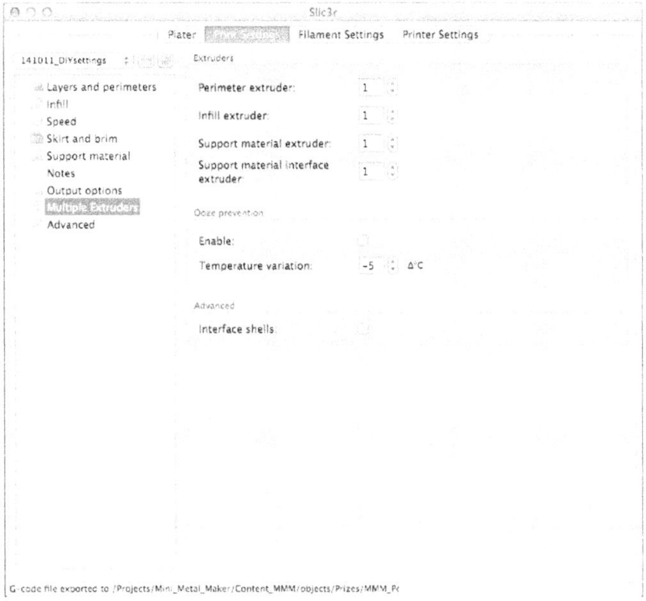

G-code file exported to /Projects/Mini_Metal_Maker/Content_MMM/objects/Prizes/MMM_Pc

Panel 4 (middle-right):

Slic3r

Plater | Print Settings | Filament Settings | Printer Settings

141011_DIYsettings

- Layers and perimeters
- Infill
- Speed
- Skirt and brim
- Support material
- Notes
- Output options
- Multiple Extruders
- Advanced

Extruders

Perimeter extruder:	1	
Infill extruder:	1	
Support material extruder:	1	
Support material interface extruder:	1	

Ooze prevention

| Enable: | | |
| Temperature variation: | -5 | Δ°C |

Advanced

| Interface shells: | | |

G-code file exported to /Projects/Mini_Metal_Maker/Content_MMM/objects/Prizes/MMM_Pc

Panel 5 (bottom-left):

Slic3r

Plater | Print Settings | Filament Settings | Printer Settings

141011_DIYsettings

- Layers and perimeters
- Infill
- Speed
- Skirt and brim
- Support material
- Notes
- Output options
- Multiple Extruders
- Advanced

Extrusion width

Default extrusion width:	0	mm or % (leave 0 for auto)
First layer:	200%	mm or % (leave 0 for default)
Perimeters:	0	mm or % (leave 0 for default)
Infill:	0	mm or % (leave 0 for default)
Solid infill:	0	mm or % (leave 0 for default)
Top solid infill:	0	mm or % (leave 0 for default)
Support material:	0	mm or % (leave 0 for default)

Flow

| Bridge flow ratio: | 1 | |

Other

| Threads: | 2 | (more speed but more memory usage) |
| Resolution: | 0 | mm |

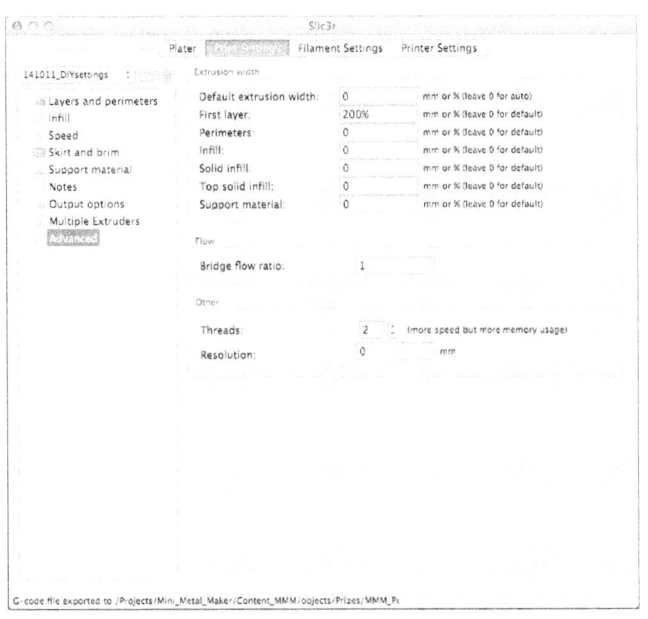

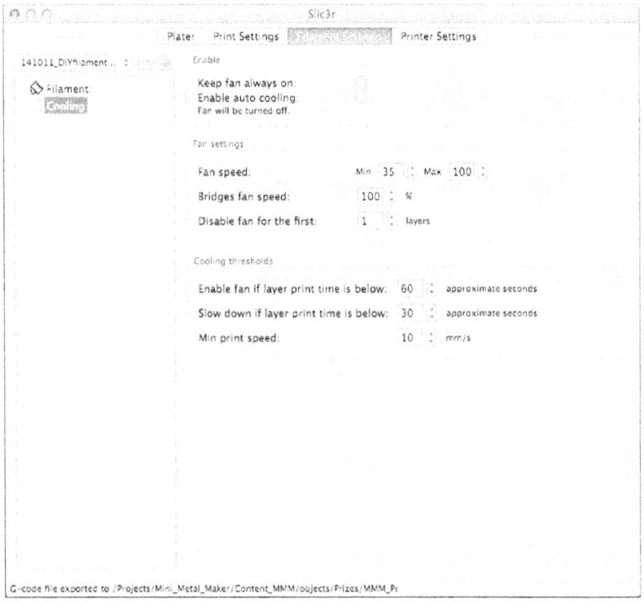

G-code file exported to /Projects/Mini_Metal_Maker/Content_MMM/objects/Prizes/MMM_Pc

Panel 6 (bottom-right):

Slic3r

Plater | Print Settings | Filament Settings | Printer Settings

141011_DIYfilament

- Filament
- Cooling

Enable

Keep fan always on:
Enable auto cooling:
Fan will be turned off.

Fan settings

Fan speed:	Min 35	Max 100
Bridges fan speed:	100	%
Disable fan for the first:	1	layers

Cooling thresholds

Enable fan if layer print time is below:	60	approximate seconds
Slow down if layer print time is below:	30	approximate seconds
Min print speed:	10	mm/s

G-code file exported to /Projects/Mini_Metal_Maker/Content_MMM/objects/Prizes/MMM_Pc

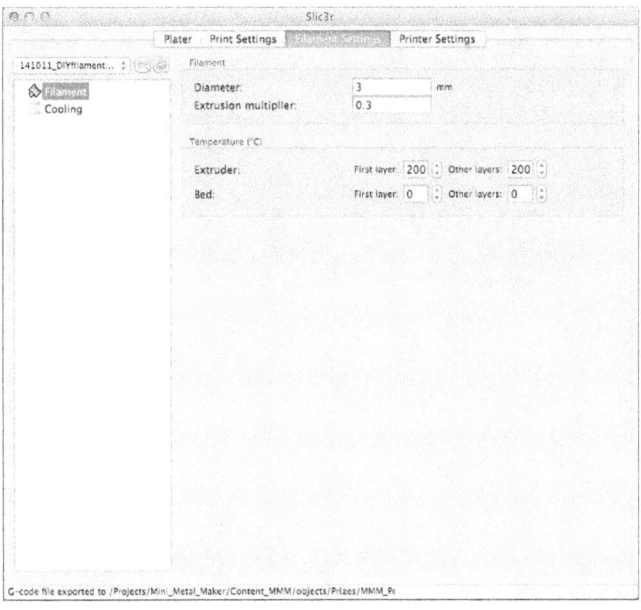

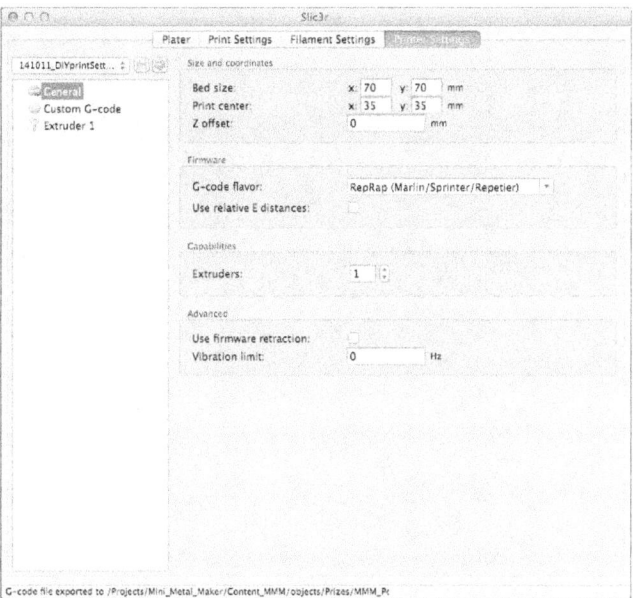

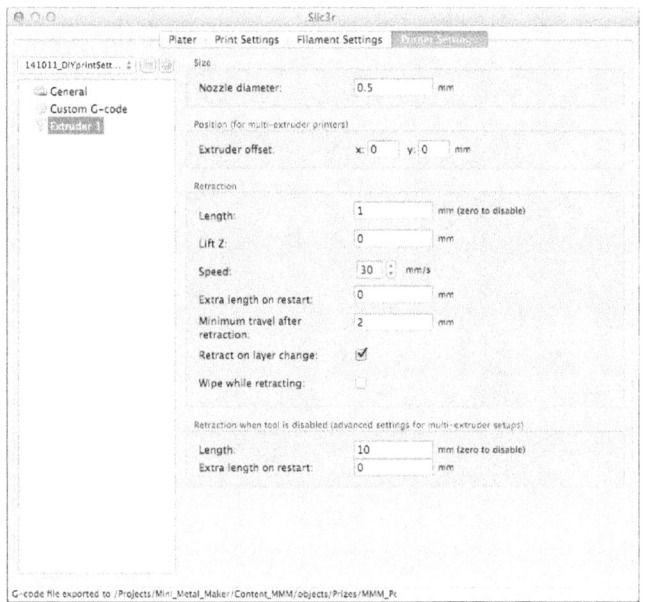

17.2 Quick links to software downloads

3D Modeling Program: tinkercad.com
http://tinkercad.com

Slicer Program: Slic3r
http://slic3r.org/download

Host Program: Printrun (Pronterface)
https://github.com/kliment/Printrun

Firmware: Marlin, modified for the Mini Metal Maker
github.com/MiniMetalMaker/MMM-DIY

Development Software: Arduino
http://arduino.cc/en/Main/Software

Mini Metal Maker

Build Your Own Mini Metal Maker
©2016 Mini Metal Maker LLC. Mini Metal Maker and Mini Metal
Maker DIY are trademarked products of Mini Metal Maker LLC.
Visit our website at www.minimetalmaker.com for more info.